A Temperate Empire

A TEMPERATE EMPIRE

Making Climate Change in Early America

ANYA ZILBERSTEIN

OXFORD
UNIVERSITY PRESS

OXFORD
UNIVERSITY PRESS

Oxford University Press is a department of the University of Oxford. It furthers the University's objective of excellence in research, scholarship, and education by publishing worldwide. Oxford is a registered trade mark of Oxford University Press in the UK and certain other countries.

Published in the United States of America by Oxford University Press
198 Madison Avenue, New York, NY 10016, United States of America.

First issued as an Oxford University Press paperback, 2019

Library of Congress Cataloging-in-Publication Data
Names: Zilberstein, Anya.
Title: A temperate empire : making climate change in early America /
Anya Zilberstein.
Description: Oxford : Oxford University Press, [2016] | Includes
bibliographical references and index.
Identifiers: LCCN 2016014091 (print) | LCCN 2016021500 (ebook) |
ISBN 9780190206598 (hardcover : alk. paper) | ISBN 9780190055516 (paperback : alk. paper) |
ISBN 9780190206604 (Updf) | ISBN 9780190206611 (Epub)
Subjects: LCSH: Climatic changes—North America—History. |
Climatic changes—United States—History.
Classification: LCC QC903.2.N665 Z55 2016 (print) | LCC QC903.2.N665 (ebook)
| DDC 304.2/5097409033—dc23
LC record available at https://lccn.loc.gov/2016014091

The Rachel Carson Center for Environment and Society and *Le Fonds de recherche du Québec–Société et culture* generously supported the publication of this book.

For my parents

Contents

Acknowledgments

I COULD NOT have researched or completed this book without the financial support and other resources provided by a *Fonds de recherche sur la société et la culture, Québec, Établissement de nouveau professeurs-chercheurs* grant, a National Endowment for the Humanities/Mellon Foundation Long Term Fellowship at the John Carter Brown (JCB) Library at Brown University, and a Rachel Carson Center Fellowship in Munich. The JCB Library and Carson Center were both extraordinary opportunities that gave me time to pursue new questions.

Although most of my dissertation did not make it into this book, the research for the former provided a basis for the latter and was generously supported by the following institutions and grant committees: Minda de Gunzburg Center for European Studies, Harvard University; Dibner Institute for the History of Science and Technology; Michael Kraus Grant in Colonial History, American Historical Association; International Council for Canadian Studies; and Center for International Studies, MIT. More important, I could not have written this book without the formative training and inspiration I received in MIT's Doctoral Program in History, Anthropology, Science, Technology and Society, especially from Harriet Ritvo, Deborah Fitzgerald, Leo Marx, Anne McCants, Peter Perdue, Stefan Helmreich, Meg Jacobs, Merritt Roe Smith, Christopher Capozzola, and David S. Jones. A special nod of appreciation goes to fellow PXYers and post-PXYers interested in environmental history: William J. Turkel, Shane Hamilton, Jenny Leigh Smith, Peter Shulman, Meg Heisinger, Candis Callison, Nicholas Buchanan, Etienne Benson, and Rebecca Woods. During a brief stint back at MIT, Mabel Chin, Margo Collett, and Chuck Munger in the History Department were exceedingly helpful. From the outset, my colleagues in the Department of History and beyond have made Concordia University a lively, interesting, and humane place to work. I am grateful for the guidance and encouragement of my chairs, Shannon McSheffrey,

Norman Ingram, and Nora Jaffary, and I thank Donna Whittaker, Darleen Robertson, and Nancie Jirků for helping to coordinate my leaves.

I was fortunate to present sections of my research to a number of audiences, and I thank the organizers for inviting me to the following seminars and workshops: *Quelques arpents de neige* Environmental History Workshop, Kingston (Daniel Rück); The Boston Environmental History Seminar, Massachusetts Historical Society (Conrad Wright); Montreal Land Workshop, Université de Montréal (François Furstenberg); John Carter Brown Library (Ted Widmer); Stokes Seminar, Dalhousie University (Justin Roberts); German Association of American Studies, Trier, Germany (Wilfried Mausbach); War and the Caribbean Workshop, University of Edinburgh (Nuala Zahedieh); Early Modern Studies Institute, Huntington Library (Ted McCormick, Vera Keller, and Peter Mancall); European Society for Environmental History Summer School, Flaran, France (Emmanuel Huertas); Rachel Carson Center (Christof Mauch and Helmuth Trischler); Occidental College History Department (Michael Gasper); The History and Politics of the Anthropocene Workshop, University of Chicago (Dipesh Chakrabarty and Fredrik Albritton Jonsson); Colonialism and Climate History Workshop, Georgetown University (Eleonora Rohland, Franz Mauelshagen, and John R. McNeill); the Empires Working Group, Princeton Institute for International and Regional Studies (Rachel Price); and the John Carter Brown Seminar in the History of the Americas and the World (Neil Safier). For permission to incorporate portions of previously published articles I thank the *William and Mary Quarterly* and the *Journal for Eighteenth-Century Studies*.

For their interest, suggestions, and help of all kinds, I am indebted to numerous people. For excellent research assistance I thank Kristoffer Archibald and Pierre-Etienne Stockland. I am grateful to cartographer Mike Reagan for drawing the elegant map of the Northeast. Joyce Chaplin and Neil Safier helped me navigate the publishing process. Joyce, three anonymous reviewers, and my editor Susan Ferber read entire drafts of the manuscript and offered shrewd advice on how to improve it. I am also grateful to Victoria Danahy for her thorough but light copyediting. For detailed reading and comments on parts of the manuscript as it developed, I thank Vicky Albritton, Danielle Bobker, Christopher L. Brown, Christopher Clark, Sarah Easterby-Smith, Carolyn Fick, François Furstenberg, Max Edelson, Michael Gasper, Jan Golinski, Michael Hill, Margaret R. Hunt, Wilson Chacko Jacob, Fredrik Albritton Jonsson, Claudia Leal, Paul Lovejoy, John R. McNeill, Heather Meek, Philip Morgan, Karen Oslund, Emily Pawley,

Susan Pennybacker, Allyson Poska, Charles Postel, Grégory Quenet, Elena Razlogova, James Rice, James Roberts, Justin Roberts, Harriet Ritvo, Leander Schneider, Emily Senior, Sverker Sörlin, Cécile Vidal, Thomas Wickman, Thomas Wien, Paul White, and Nuala Zahedieh.

Many thanks to Harriet Ritvo for her insight, support, and example, which have influenced my way of thinking in more ways than I can reckon. Friends offered their unwavering confidence in me and provided refreshment of various kinds, which I needed very much. Even with all of that fortification, I was not always in a temperate state of being; thank you Michael Gasper, Omri Moses, Rachel Price, and Gavin Taylor for suffering it more than most and helping to steady me. For weathering the worst storms, my family deserves much more than just thanks, especially my parents, who bravely transported us from another northern place.

A Temperate Empire

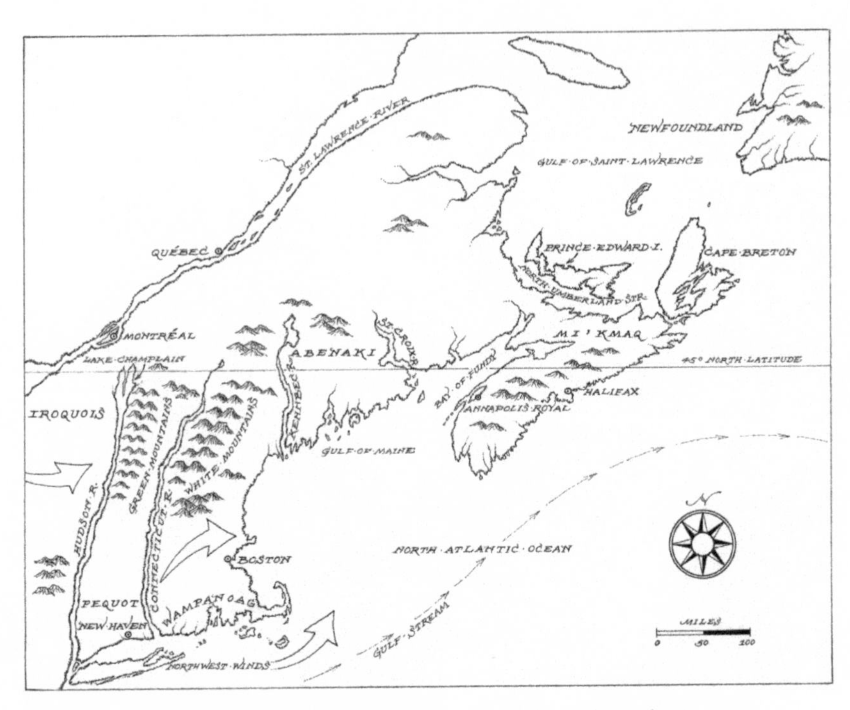

FIGURE 0.1 A Map of the Northeast. Original map drawn by Mike Reagan.

Introduction

IMPROVING THE CLIMATE

The whole earth is less subject to extreme cold than it was formerly: And that every climate is become more temperate, and uniform, and equal: and that this will continue to be the case so long as diligence, industry, and agriculture shall mark the conduct of mankind.

SAMUEL WILLIAMS, "Change of Climate in North America and Europe" (1789)

We shall probably be told, that so far as regards the physical *maladies of climate, we are to* hope *that in* process of time, *when the atmosphere of these regions shall be more impregnated with phlogistic particles from myriads of reeking dunghills, from the fumes of furnaces, from the fires and smoke of ten thousand crowded cities* hereafter to be built, *and by a general subjection of the soil to agriculture, carried on to the Arctic Circle, they may be considerably alleviated.*

EDWARD LONG, "A Free and Candid View of a Tract, Entitled 'Observations on the Commerce of the American States'" (1784)

Since greenhouse gases are chiefly the result of human industry and agriculture, it is not an exaggeration to say that civilization itself is the ultimate cause of global warming.

ANDREW REVKIN, *Discover* (1988)

WHEN EUROPEAN COLONISTS settled in North America, they improved the climate and made it temperate. From Nova Scotia to Florida, colonial farmers changed the weather, reducing the length and intensity of the North's bitterly cold winters and of the South's sweltering summers. By the late eighteenth century, "every climate" in the settler regions of British North

America and the United States had become "more temperate, equal, and mild" compared with the climates in the experiences of previous generations.

Or so thought the eighteenth-century naturalist Samuel Williams, who studied the early American climate by combining historical sources with his and other Americans' records of local weather and temperature measurements. As Hollis Professor of Mathematics and Natural Philosophy, he recorded daily air temperatures in the vicinity of Harvard College from 1780 to 1788. As a corresponding member of the Palatine Meteorological Society, an international network of weather observers based in southern Germany, Williams traveled from Cambridge to Springfield, Massachusetts, and from New Haven, Connecticut, to Burlington, Vermont, plunging a thermometer into well waters wherever he stayed overnight. Deep well-water temperature did not vary much with the seasons, so from these observations he believed he could derive average annual air temperatures across the region. In another experiment, he compared daily ground temperatures in a pasture and a shady forest in Rutland, Vermont, over the course of a growing season. He also tracked bird migrations and the budding, flowering, fruiting, and harvest dates of common native and imported trees, shrubs, and crops like apples, gooseberries, and maize. Compiling all of this quantitative data, he correlated it to a collection of historical anecdotes about New England's weather going back to the early seventeenth century.

In 1789, he drafted "Change of Climate in North America and Europe," in which he came to this conclusion: Over 150 years of permanent European settlement and land development had tempered New England's formerly severe climate. This local shift was only the latest instance of a widespread transformation that aligned the history of the Northeast and other settlements in North America with European countries in the temperate zone. Seen from a very long-term perspective, both Providence and geological processes had made "the whole earth . . . less subject to extreme cold than it was formerly," but the last hundred or so years of change visible across the Atlantic world was the result of concerted human effort, rather than divine or natural causes. Going further, he predicted "that every climate is become more temperate, and uniform, and equal: and that this will continue to be the case so long as diligence, industry, and agriculture shall mark the conduct of mankind." Productivity promised to remake the global climate for the better.

He shared these findings in a private letter to Sir Joseph Banks, president of the Royal Society of London for Improving Natural Knowledge and the most powerful patron of science in the Anglophone world. But Williams remained cautious about going "public on the subject" of human intervention in the climate. Instead of publishing his work right away, he decided to continue gathering data. In 1794, he included an account of anthropogenic climate change in his book *The Natural and Civil History of Vermont*, in which he explained that local changes in the land created reciprocal changes in the air. Since settlers began cultivating their properties in this new state, "the cold of the winters decrease; the rivers are not frozen so soon, so thick, or so long, as they formerly were; and the effects of extreme cold, in every respect appear to be diminished." He was fairly certain that this warming trend signaled "a permanent alteration."[1]

Two centuries later, modern climate scientists reached a different conclusion: The climate of the colonial Northeast had been subject to global cooling rather than warming. Using proxy data derived from ice core samples, coral, tree rings, volcanic deposits, and the like, scientists reconstructed historical weather patterns across the early modern North Atlantic world. Their computer models showed that, from roughly the fourteenth through the nineteenth centuries, the Northern Hemisphere was subject to protracted winters and abbreviated, cool, and wet summers characteristic of "the Little Ice Age" (a term coined in 1939 by American glaciologist François Matthes to describe significantly colder time spans between proper ice ages in Earth history). Over the long period of the Little Ice Age, expanding glaciers and sea ice caused anomalous or unusually severe weather around the world; across the Northern Hemisphere, there were subtle but significant drops in annual temperatures, especially in coastal regions on both sides of the North Atlantic Ocean. During peaks in this comparatively colder period, changing climatic conditions sometimes exacerbated or otherwise played a role in shaping colonial encounters. In the seventeenth through early eighteenth centuries, Native American groups in the Northeast took advantage of the prevailing hard winters by developing strategies for repelling colonists' territorial advances. The Pennacook, Androscoggin, Kennebec, Penobscot, Maliseet, and Mi'kmaq, for instance, commanded the formidable northern uplands around the Gulf of Maine, maintaining a buffer against northeastward-bound settler encroachment. Colonial farmers were especially vulnerable to late spring or early autumn killing frosts, or both, such as occurred in 1816. In this so-called year without a summer—caused by the coincidence of waning Little Ice Age conditions and the reverberating effects of the eruption of distant Mount Tambora—intense, prolonged winter weather brought harvest shortfalls, higher food prices, and hunger across the region.[2]

But early Americans' perceptions of climate history did not always match the trends twenty-first-century climatologists have identified in the past. This divergence is dramatically clear in Williams's argument for climate warming in North America, which runs in opposition to sound modern research. Even as people in the seventeenth and eighteenth centuries experienced and responded to cycles of harsher weather, the same kind of precise technical information about long-range, large-scale climatic patterns like the Little Ice Age was unavailable and unknown to them. And yet, if material evidence and computer models convincingly show that average annual temperatures in the Northeast were significantly colder even in the last decades of the Little Ice Age—from the second half of the eighteenth through early nineteenth centuries—how could Williams believe that colonists had made the weather milder?[3]

Williams was not a lone crank in his conjectures about anthropogenic climate change. Nor, despite his cautiousness, was he hazarding a new or radical theory. Instead, he joined many other learned elites who contended that European farming or settler colonialism induced a process of climate change toward a temperate ideal. Williams's conclusions corroborated those of naturalists who studied the climate history of the Lower South, the Chesapeake, the Middle Colonies, Canada, West Africa, Scotland, and the Swiss Alps; they too noted a similar form of moderation in the seventeenth and eighteenth centuries. The belief in amelioration was particularly widespread in early America and across the British Empire, where speculations about the mutual influences between political history and climate history became a standard part of transatlantic scientific debates. In 1763, Benjamin Franklin endorsed "the Truth of the common Opinion, that the Winters in America are grown milder." A few years later, the physician Hugh Williamson told an audience gathered in Philadelphia at the American Philosophical Society that elderly residents of Pennsylvania and neighboring colonies noticed "a very observable Change of Climate," namely "that our winters are not so intensely cold, nor our summers so disagreeably warm as they have been" four to five decades ago. In his *Notes on the State of Virginia*, Thomas Jefferson reported his impression that the local climate had "become much more moderate within the memory even of the middle aged." In places like Florida and Nova Scotia, where the settler population was just beginning to grow, local elites expected these changes to transpire in the near future.[4]

Besides providing empirical and anecdotal evidence of climatic improvement, early Americans also drew on classical ideas proposed by ancient historians, who had interlinked climate history with imperial history. In antiquity, writers argued that ascendant societies inhabited or created temperate environments with the "best mixture of conditions": fertile soils, a diversity of wild and domesticated species and, above all, moderate seasonal temperatures, sunlight, and precipitation. These conditions not only corresponded with healthy, large, and prosperous populations but also supposedly resulted in higher stages of cultural development, which in turn provided a stable foundation for territorial expansion. If the Mediterranean climate explained the flourishing state of imperial centers, Greek or Roman invasion of outlands would progressively soften their harsh climates and the allegedly barbarous, politically fragmented societies peopling them. Where conquerors cleared dense forests and planted deserts, they beneficially reduced excessively hot, cold, arid, or humid conditions. The decline and contraction of empire accordingly triggered regression.[5]

American colonists reinvigorated this idea and charged it with new meaning by making the domination, creation, and conservation of temperate climates into an abiding political problem. For agents of the English (after 1707, British) Empire intent on settling colder regions like northeastern North America, the climate was acutely problematic. Ptolemaic tradition dictated that the burning heats at the equator and constant frosts at the poles formed the temperate zone's fixed boundaries around the world, so the seasons in the area between the Gulf of Saint Lawrence and Long Island Sound should have been more or less the same as those on the coast of the northern Mediterranean Sea. Instead, the Northeast's winters and summers defied early explorers' expectations.[6] Surrounded by teeming cod and whale fisheries, hardwood and coniferous forests, and the interior fur trade, the region that eventually became known as New England and Nova Scotia also formed the northeastern limit of arable land on the continent and the southernmost extent of the ice sheets before their retreat about 10,000 years ago. Despite hilly topography, soils heavily intermixed with glacial debris, and nearly subarctic winters, the English nevertheless committed themselves to conquering the indigenous territories of the Northeast and resettling them with Europeans.[7]

After struggling to survive in the region's environment, several groups abandoned their early colonial projects altogether. According to Samuel Champlain, Portuguese fishermen relinquished plans to colonize Cape Breton Island after passing one winter there. The same was true of the Virginia Company of Plymouth's 1607 Sagadahoc settlement on the Maine coast and the Scots' short-lived attempt in the late 1620s to establish a colony in the area that heartier French colonists soon renamed Acadia. These failures contributed to an emerging perception of the Northeast as an irredeemably cold desert, prompting ongoing transatlantic debates about whether or not it was suitable for permanent European settlement and agricultural development. In surveys of the climate across the Americas, writers often described the weather in New England and Nova Scotia as comparatively harsh, excessive, extreme, intense, rigorous, bitter, and severe. These and similar adjectives recurred in numerous publications and private communications that circulated in the Atlantic world from the early modern period through the turn of the eighteenth century. By the beginning of the nineteenth century, although some believed that the Northeast's environment was improving, others continued to have an "oblique and unworthy opinion" of it. As Yale College President Timothy Dwight complained, the northern "climate is supposed" by outsiders "to be inhospitable, and the soil barren." Migrants unwilling to tolerate numbing cold or to tackle the challenge of farming in the region usually avoided it or, if they could, quickly departed for warmer places. However, for anyone interested in encouraging immigration to and developing property in the region, pessimistic accounts of the local climate became a persistent source of anxiety. Worried about the political and economic consequences of the climate's bad reputation, local elites became intensely interested in determining the true character of the local climate.[8]

To better understand northeastern North America's climate, learned elites studied its natural history. Methodologies for researching climate, however, were relatively limited, indirect, and subjective compared with other aspects of the natural world. Although naturalists contained, preserved, and transported specimens of minerals, plants, and creatures in order to observe and compare them, they could not so easily isolate, objectify, and circulate meteorological or atmospheric phenomena, let alone climatic patterns as a whole, which unfolded on a variety of not entirely commensurable temporal and geographical scales. Instead, they developed qualitative and quantitative proxies for studying the natural history of climate. These records included descriptive reports, weather diaries, and instrumental measurements of temperature, pressure, humidity, and wind speed using thermometers, barometers, and other devices of varying reliability. By the eighteenth century, a number of privileged men in the region possessed these instruments, but qualitative proxies remained the most plentiful because the climate had always been an element of the colonial environment itemized and described in official correspondence or popular print culture, such as settlement schemes, local and natural histories, and gazetteers. After the mid-seventeenth-century establishment of the Royal Society and the Lords of Trade and Plantations (later the Board of Trade), these observations often appeared at the beginning of such texts, following a template made explicit in founding member of the Royal Society Robert Boyle's instructions that descriptions of the air, climate, and temperature should be accorded first and second place in writing "the natural history of a country great or small."[9]

Naturalists also tried to understand climate in terms of what would later be called the science of acclimatization. In its most elementary form acclimatization was the practice of moving a living thing far from home and observing how it fared in a new place. To predict survival rates, acclimatizers extrapolated the similarities between native and foreign habitats from accounts of long-distance migrations of people, animals, and plants and their success or failure in adapting to unfamiliar environments in both the near and the very remote past. In the seventeenth century, emerging theories about the deep history of the Earth suggested that climatic changes might have stimulated primeval migrations. Fossils and preserved skeletons of extinct creatures found in northern latitudes that resembled living tropical species offered the most compelling material evidence. By far the most sophisticated, influential theorist to incorporate such evidence was George-Louis Leclerc, comte de Buffon, who asserted that natural historians, like other scholars, had to research "the archives of the world." In *Des époques de la nature* (1778), the volume on geology in his widely read *Histoire Naturelle*, Buffon proposed that all life on Earth first emerged in the Arctic many thousands of years ago when it had enjoyed a much warmer climate and the rest of the world was too hot for habitation. As the global climate cooled, plants, animals, and people gradually drifted southward and dispersed among the continents.[10] Since the fifteenth century, European empires further enabled this ongoing process of biogeographical shuffling by encouraging or forcing transoceanic migration and ecological exchange on an unprecedented scale. Of course transatlantic colonization and the slave trade were not conceived as experiments, but for natural historians the so-called seasoning of people and acclimatization of other species when they moved to unfamiliar environments offered clues about the climate in faraway places. Deriving knowledge about local climates from such clues proved to be a highly inexact, controversial science. It was especially equivocal in its contribution to speculations about the connection between climate and cultural diversity, including increasingly strong beliefs about the immutable relationship between climatic geography and racial difference. As a result, even as descriptions, measurements, and theory about the global climate and its history proliferated in the seventeenth and eighteenth centuries, reliable knowledge about the particular environmental history and characteristics of local climates remained tentative.[11]

A Temperate Empire examines the range of these often-conflicting ways in which metropolitan and local elites—officials, large landowners, land speculators, merchants, ministers, naturalists, and agricultural improvers—attempted to come to terms with the climate of northeastern North America, including their engagement in scientific debates about the real and perceived limits a cold northern climate posed to settler colonialism. Colonizing and inhabiting the Northeast required understanding—and perhaps even changing—its climate. Because no one knew for certain if dramatic or lasting physical climate change had materialized, early Americans were most effective at destabilizing ideas about the local climate. They made ideas about the climate change by writing about it. In the process of describing places like northeastern North America that were not as temperate as first assumed, local elites stretched the imagined geography of the temperate zone, shifted their definitions of bodily comfort, and projected a more temperate future for the region, once they had succeeded in attracting more permanent, industrious settlers such as themselves.

Europeans did not simply discover the timeless climates of the Western Hemisphere. Realizing that the ancient equation of climate with latitude was inaccurate for the Americas, they studied and tried to remake local climates according to the imperatives that informed their plans for colonial settlement, population growth, and economic development. By focusing on these political, intellectual, and rhetorical dimensions of early discussions about the region's weather and climate, this book argues for the necessarily dynamic, historical nature of both climate knowledge and of our contemporary debates about it.

In a sense, climate has long played a part in the origins of American history. Considerations about the climate—mainly stories of frostbitten Pilgrims and Puritans—form the backdrop to classic scenes of colonial New England. The earliest chroniclers of Plymouth Plantation and Massachusetts Bay Colony presented the violent conquest and settlement of indigenous territory as a history of precarious survival in a cold, "hideous and desolate" wilderness. Because these men led groups of separatists seeking religious refuge from the "Intemperance" of the Old World, their winter tales carried both worldly and scriptural meanings. They were parables within a parable. In the larger, ongoing narrative of western expansion, the English colonization of northeastern North America was a new chapter that began with a trial by frost. In the microcosm of New England, God challenged the faith of Protestant congregations with the hazards of a bewildering environment: the harsh elements, the potential threat of hostile Indians, and the risk of spiritual disorientation. Their miraculous triumph was the result of Providence—and perseverance. For any migrant, adjusting to a new environment could be harrowing. For pious Calvinists, difficulties and narrowly averted dangers were accepted dimensions of everyday life. Their willingness to overcome hardship was another measure of their steadfast faith. Before embarking from Leiden on the *Mayflower*, two men wrote to reassure their Virginia Company sponsors that "it is not with us as with other men, whom small things can discourage, or small discontentments cause to wish themselves at home again." The settlers patiently endured "all the misery that Desart could put upon them" and used "their wits to make their best use of that then Snow-covered land for their necessities."[12]

The earliest English migrants to the region portrayed themselves as vulnerable, devoted settlers rather than as brazen conquerors. They were merely humble pioneers who accepted that "all beginnings are ever difficult." Until the late twentieth century, most historians reinforced this view, using the colonization of New England as a restrained but victorious origin story from the United States' prenational era. Textbook accounts emphasizing the religious beliefs and farming communities of New England typically downplayed the misfortunes of the Virginia Company and the miseries of laborers in the Chesapeake. Christian piety and small proprietorship—not corporate capitalism and plantation slavery—were the basis of American society. For other writers, New England served as the archetype of modern society and the tendencies associated with it. For Max Weber, Perry Miller, and Immanuel Wallerstein, religious doctrine strongly influenced commercial behavior in the northern colonies. The region's ministers and devout merchants were paragons of productive behavior. That they "employed their wits" to make the cold desert into a home was a mark of their rationality and disciplined asceticism. The jeremiads of the late seventeenth century expressed a pervasive but ultimately productive anxiety of failure that propelled northerners to industriousness in their daily lives. Their aggressive work ethic and resourceful efficiency explained how and why this colonial periphery was successfully converted into a worldly metropolitan center of industrialization.[13] Combining these perspectives, social historians showed that, for all the moral economy engendered within New England and Nova Scotia's small, insular, churchy communities, householders were both producers and avid consumers, engaged in and reliant on exchanges across the Atlantic world.[14] Building on this consumerist argument, environmental historians cast northern colonists as rapacious and wasteful. Their utilitarian approach to American nature as an abundant economic resource was the basis for the region's industrialization but also its concomitant environmental decline. That the secular transformation of agrarian to industrial capitalism emerged from a community originally composed of religious refugees was not as paradoxical, or as unequivocally admirable, as it might have originally appeared.[15]

In most of these historical accounts, the climate figures, if at all, as an atmospheric narrative element: Cold winters provide a vivid challenge to the early stages of colonization that ultimately prove to be an incidental hurdle to settlement. Or, less dramatically, colonial farmers eventually incorporate considerations about the seasons and other local environmental conditions into their agricultural practices and economic calculations. This neglect is partly the result of assuming—implicitly or explicitly—that the Northeast was essentially the same as Britain and northern Europe, taking for granted the stable natural order of climate zones. Accordingly, most historians of colonial climates have examined the creation of neo-European landscapes in temperate settler colonies—a process Alfred Crosby influentially named "ecological imperialism"—and the limits to European settlement in hot environments, particularly the difficulties colonizers faced in assimilating to tropical disease ecologies in the Caribbean, Africa, and South Asia. Until recently, however, climatic instability—and early Americans' alertness to it—has rarely served as a focus of sustained historical analysis.[16]

Although concentrating on a particular region of British North America, the chapters that follow examine how people conceived of climatic instability and variability in the past. Theories about acclimatization, environmental determinism, and anthropogenic influences constituted the most prevalent understandings of and discussions about climate and climate change in the early modern period through the early nineteenth century. Learned elites in New England and Nova Scotia gave local expression to these early ideas about climate by applying them to colonial policies to encourage immigration, settler population growth, and economic development. They argued that the regional climate was not inescapably harsh. It was sufficiently temperate, it was becoming temperate, or—in the least- or last-populated sections—it would eventually become perfectly temperate through active land improvement.

The first part of the book explores the geography of these arguments. Chapter 1 asks how the colonization of northeastern North America revised the ways in which Europeans imagined the global geography of temperate, tropical, and polar climates. Two aspects of New England's and Nova Scotia's history complicated attempts to understand whether or not its climate was in the temperate zone: 1) the region's surprisingly severe and changeable weather and, 2) its unstable colonial boundaries in the seventeenth through the early nineteenth centuries. Colonization revealed the inadequacy of ancient models of the world's climate zones based on latitude. As a result, in the New World generally and in the Northeast in particular, knowledge about the ecological boundaries of the temperate zone was highly imprecise. However, because no new cartographical representations of the climate were produced in this period, the region's biogeography was subject to ongoing dispute and competing textual descriptions, debates that eventually contributed to a more expansive, labile definition of what counted as a temperate climate and where one could be found.

Chapter 2 turns to the geography of early climate science in New England, Nova Scotia, and the broader Atlantic world by asking who had the authority to produce knowledge about local climates. In the eighteenth century, learned men and women in the region began to emphasize empiricism over theory, arguing that the collection of firsthand observations and measurements was the only way to refine knowledge about the northern climate. By reconstructing the social networks that served as the primary channels to communicate and assess the accuracy of this empirical information, the chapter shows that climate knowledge was produced as a result of engagement in transatlantic scientific debates. Before the rise of professional science and its institutions later in the nineteenth century, correspondence networks and other forms of cosmopolitan sociability constituted scientific culture. Early American naturalists, like their counterparts elsewhere, were governing elites and learned gentlemen and women who, in addition to their primary occupations or other pursuits, also studied, described, and managed local environments. By participating in debates about the local climate and other aspects of the natural world, learned elites in the Northeast established their credibility as naturalists. The crucial contacts between British and North American naturalists before the American Revolution, moreover, were not entirely disrupted by it. Naturalists in various locales remained connected through these circuits, which persisted through the early nineteenth century as part of the increasingly transnational scientific community of the wider Atlantic world. On all sides of new political loyalties and boundaries, naturalists continued to share studies of—and debate answers to unresolved questions about—the northern climate.

The three chapters in the second part of the book examine the politics of climate in New England and Nova Scotia, particularly tensions between local and distant elites about how to mobilize climate knowledge in the service of colonial settlement and economic expansion. Chapter 3 focuses on debates about the possibilities and limits of northern acclimatization as the basis for increasing the settler population and developing agriculture in the region. It recounts how British officials—particularly in areas that bordered Catholic New France—manipulated perceptions of the regional climate in attempts to increase the Protestant settler population and diversify the economy by introducing frost-tolerant silkworms and wine grapes, among other schemes. Beginning with the first British governor of Nova Scotia, Samuel Vetch, who survived Scotland's disastrous colonial experiment at Darien in Central America in the 1690s, over the course of the eighteenth century a series of local elites conjured a transatlantic northern temperate zone so as to persuade supposedly cold-hardened migrants from the British Isles and northern Europe to permanently resettle in the Northeast. Acclimatizers acknowledged the coldness of the northern climate but argued that it was suitable for those humans, animals, and plants that were already—or that would quickly become—desensitized to it. As British official and Vetch's collaborator Francis Nicholson put it to Queen Anne in 1710, the climate of the northern colonies was "much more callculate" to our "Northerne constitution than Darien." Successive projectors similarly combined climatic determinism with political arithmetic, favorably comparing the region with northern England, Scotland, Scandinavia, and parts of Russia and China, all more-or-less civilized areas "in the Northern temperate zone." In most cases, northern acclimatization proved to be irresistible as policy rhetoric despite the fact that it was a repeatedly ineffective strategy for recruiting loyal Protestant migrants.[17]

Other northern elites firmly believed they inhabited a temperate climate rather than a far northern fringe of tenuous settlement. These extreme-climate deniers brushed away complaints about New England's and Nova Scotia's unbearably cold winters, muggy summers, heavy fogs, strong winds, or the climate's changeability. William Wood remarked in the early seventeenth century that the region's sometimes erratic weather simply made manifest the "axiome in Nature" that everything changes, "*nullum vsole temst perpetuum*, no extreames last long." Chapter 4 relates the most vivid example of this rhetorical approach to minimizing the inconveniences or hazards of the climate: the controversy over Nova Scotia Lieutenant Governor (and former governor of New Hampshire) John Wentworth's frustrated quest between 1796 and 1800 to settle almost 600 Jamaican Maroon deportees in the province, in spite of their and their abolitionist advocates' objections that the winters were too severe for them. If Vetch had argued that Europeans were supposed to arrive preconditioned to withstand the region's cooler climate, Wentworth insisted that Africans transported from hot climates might be tempered by it. In particular, he exploited medical advice about the degenerative effects of excessively hot climates and the remedial effects of the local climate in formulating a settlement policy for the Maroons that doubled as a social reform. This was another case involving traffic between the Caribbean and the Northeast, which shows how the emergence of racial ideology and the abolitionist movement intersected with political and scientific debates about the relative merits of inhabiting a northern climate.

Finally, Chapter 5 explains why increasing the permanent settler population was thought to be so crucial to northern development. It examines the widespread belief that landowning farmers might eventually temper the climate as long as they could be persuaded to continuously improve their properties—that is, to responsibly exploit and develop local natural resources—as learned elites saw fit. The chapter argues that transatlantic debate about the likelihood of anthropogenic climate change in the seventeenth and eighteenth centuries was inextricably linked to and can only be adequately understood in terms of the broader preoccupation with agricultural improvement. In New England and Nova Scotia, local elites harshly criticized their neighbors' farming practices as outdated and offered unsolicited advice about how to adapt the latest scientific agriculture to the region. Nearly all American and British elites thought that the introduction of new crop rotations, greenhouses, and livestock breeding techniques would increase the productivity and profit of individual farms in the short run. But Samuel Williams and many others also believed these were prescriptions for ameliorating the harshness of the climate in the long run. From this optimistic point of view, the region's seemingly more temperate climate at the end of the eighteenth century was an artifact of agricultural improvement. Whether they were correct—either in the eyes of contemporaries who were skeptical of anthropogenic climate change, like Jamaican planter Edward Long, or in hindsight—these learned elites articulated an early scientific argument that linked increases in population and industrious activity with radical transformations in weather and climate.[18]

Although many aspects of eighteenth-century theories about acclimatization, climatic determinism, and climatic improvement are largely outdated, elements lingered on well into the next centuries. In the arid American West, foresters assured land-hungry migrants that much-needed rain and a more equable climate "would follow the plow." Latitude continued to inform popular notions about climatic geography, including the suspicion that black people did not really belong in the North. Northern countries, "must ever be . . . inhabited by the descendants of Northern races," urged a founder of the nationalist Canada First movement in the 1860s. Some dreamed that a racial population policy would hasten, as one popular early twentieth-century title had it, "the Northward Course of Empire." And white Australian settlers, anxious about their far remove from the centers of modern capital, became intent on proving that their continent mirrored the temperate geography of the Northern Hemisphere.[19] In northeasteastern North America, the technological and commercial successes of the Industrial Revolution overshadowed older anxieties about the ways in which the region's climate hindered population growth and economic expansion. The rapid and visible industrialization of manufacturing in the region encouraged an about-face. The North, long perceived as a place of limited potential, became newly associated with dynamism and progress, obscuring the agonized attempts over two centuries to prove that it was always so. The images of a cool, temperate North and a balmy, dissolute South came to inform ideas about the modern geography of progress and backwardness. These dichotomies and their attendant clichés accumulated meaning in the national contexts of the nineteenth-century United States and Canada as well as across the British Empire, contributing to some of the strongest, most deplorable articulations of racial theories of human difference. In the second half of the twentieth century, the stereotypes about temperate versus severe climates took on a much broader resonance in assumptions about the contemporary geography of economic development, underdevelopment, and the horizons of possibility, captured in the monolithic terms Global North and Global South.

More recently, actual climate change is intensifying some and undoing other aspects of this imagined global order, once more shifting the geography and definition of what counts as a temperate climate as opposed to one too extreme for human livelihood. The Inuit and other northern nations are scrambling to lay claim to an ice-free Arctic at the same time as warming polar conditions are forcing a range of species migrations. Rising oceans are submerging tropical islands, prompting discussions about the extent to which people are equipped to adapt to climate change and generating a new category of exile, the "climate refugee." The scientific consensus is that these and many other developments are due largely to accumulated human economic activities, a concept that has generated public alarm as well as bitter disputes about whether and how to respond. The idea of anthropogenic climate change has faced inordinate resistance in recent decades, taking an especially long time to gain political traction in the United States, despite the urgency and unequal international distribution of its real threats.[20]

In part because of the overwhelming magnitude of this global emergency, it may be tempting to think that such political debates about climate, climate change, and climate science are wholly unprecedented. In the midst of the accelerating planetary crisis of global warming, however, it is instructive to consider the implications of earlier, ubiquitous debates about the historical relationship between people and the climate. From the seventeenth through the early nineteenth centuries, prominent figures in North America and Europe, many of them at the highest levels of government, routinely insisted that considerations about the climate and climate change were inseparable from the management of natural and human resources. They openly acknowledged that scientific understanding of the complex causes and consequences of environmental change—including the historical, present, and future state of the climate—was necessarily provisional. And yet, in this preindustrial period, the verdict was that people and their economic activities were undoubtedly capable of being agents of such changes, including their power to transform the climate.[21]

Moreover, for early Americans, climate change was an intended consequence of economic development. They felt encouraged by the beneficial moderation of extreme climates, which they understood as a deliberate, righteous, and anxiously awaited outcome of population growth and industrious activity rather than their unintended, lamentable, uncontrollable, and only belatedly recognized side effect. Contemporary engagement with global warming is thus in some ways a dramatic reversal of the early modern prophesies discussed in this book. In another sense, the current predicament is their fulfillment, especially now that the current geological era, marked by the inextricability of human and natural history, is already being called the Anthropocene, evoking humanity's collective agency in changing the Earth.

As should be plain from the continuities in colonial debates about climate that this book explores, conceiving of climate change, even on a planetary scale over the course of millennia, is not so novel. Various ways of thinking about climate, including anthropogenic climate change as part of a progressive ideology about the improvement of nature, are very old habits of thought. What is new is the dire necessity of reversing our understanding of their environmental and social repercussions. By pointing to these kinds of ironies, *A Temperate Empire* demonstrates the ineluctable ways in which political disputes, economic interests, and cultural values mediated ideas, interpretations, and responses to weather and climate in the past. The climate has long been an ambiguous, fraught, and contested sign of historical change.

PART I

Climate and Geography

I

The Golden Mean

I hope you effected your journey with health and pleasure, and found your family as well as the frigid sensations of a Polar climate can admit to its inhabitants. I have often wondered that any one should settle in a cold country while there is room in a warm one; and lamented that yourself and Dr. Priestley should have been led into the snows of Maine and Northumberland rather than the genial climates of the South.

THOMAS JEFFERSON TO BENJAMIN VAUGHAN

EUROPEAN TRAVELERS, ESPECIALLY migrants who intended to resettle in American colonies, usually hoped to avoid the perpetual summer of the tropics and the never-ending winter of the Arctic. These extreme climate zones were plainly "unfit for habitation." Instead, settlers sought regions with a mild climate divided into four seasons of equal duration, which promised to moderate both the temperature of the air and the temperament of anyone breathing it in. "Creatures that participate of heate and cold in a meane, are best and holsomest," wrote Massachusetts colonist Thomas Morton in the early seventeenth century, paraphrasing Aristotle. With less comfort and more struggle, people might live anywhere, but "the middell Zone betweene the two extreames is best, and it is therefore called Zona temperata, and is in the golden meane, and all those Landes lying under that Zone, most requisite and fitt for habitation." William James Almon, a Rhode Island–born physician who moved to Nova Scotia during the American Revolution, also seized on this idea when he copied a passage from Adam Ferguson's *Essay on the History of Civil Society* into his notebook: "Man, in his animal capacity is qualified to subsist in every Climate. He reigns with the Lyon and the Tyger under the equatorial heats of the sun, or he associates with the Bear and the Reindeer beyond the polar circle. . . . The intermediate climates, however, appear most to favor his Nature."[1]

There was general agreement on the benefits of occupying territories in the temperate zone. Yet from initial conquests and over the course of two subsequent centuries there was no definitive knowledge about its geographical extent in the Western Hemisphere. Was northeastern North America located within the habitable temperate zone? As Europeans colonized and amassed information about the region, responses to this deceptively simple question became increasingly contested. According to the oldest geographical understanding of climate zones organized by latitude, the answer was unequivocal: Temperatures, sunlight, and precipitation on the coast of North America between 40° and 48° North latitude would be identical to conditions on the northern Mediterranean littoral. But colonial experiences with the Northeast's formidable winters and short, sultry summers revealed a puzzling asymmetry between climate and latitude. According to decades of personal observations and adverse publicity about the region, the answer was no. Throughout the sixteenth to early nineteenth centuries, a bleak vision of the region competed with its antithesis. Many writers produced glowing reports of a rich, temperate environment; other accounts emphasized its rocky soils and climatic extremes.[2]

In the first published account of the region, *A Briefe and True Relation of the Discoverie of the North Part of Virginia* (1602), John Brereton wrote "truely" of "the holsomnesse and temperature" of its climate. Samuel Champlain and Pierre Biard declared the climates of New France and Acadia to be as "salubrious" as those of France, with temperatures comparable to those in Grenoble, Vienna, and Bayonne. Captain John Smith's 1616 *A Description of New England* extolled the "moderate temper" of the air. New England was "well inhabited with a goodly, strong and well proportioned people" and, of the twenty-five seamen in his crew, none had gotten sick while visiting the coast. The region's healthy air, abundant forests and fisheries, and the "excellent soyle" in which Native Americans planted maize and other crops—these many natural advantages offered "proofe of an excellent temper." "Who can but approove this a most excellent place, both for health & fertility? And of all the foure parts of the world that I have yet seene not inhabited," Smith enthused, blithely disregarding his descriptions of Native villages, "could I have but means to transporte a Colonie, I would rather live here then anywhere."

Because America "never came to the knowledge of any Hebrew, Greek, or Roman," the Ancients decided that "these zones could not be inhabited." "Now," wrote the Scottish gentleman William Alexander in 1624, Europeans knew that the Western Hemisphere contained some of "the most pleasant parts in the world," including the land between the Bay of Fundy and the Gulf of St. Lawrence, where, though he had yet to visit, he hoped to establish a New Scotland. William Wood and Thomas Morton, who wrote accounts of New England in the 1630s, agreed with him. In New English Canaan, as Morton called the region, no colonist was "ever knowne to be troubled with a cold, a cough." It boasted "for certaine the best ground and sweetest Climate in all those parts" of America. Given all of these advantages, it was clear—at least to these early colonists—that "the wise Creator of the universall globe" had situated the region in "the temperate Zone."[3]

Having read John Smith's account, William Bradford and the hundred or so migrants traveling with him on the *Mayflower* might have looked forward to landing in a place "as temperate and as fruitful as any other parallel in the world."[□] But in the early winter of 1620, when they approached northeastern North America, "the weather was very cold." It was so bitter, Bradford recounted in his history of Plymouth, sea spray "froze hard." People standing above deck looked "as if they had been glassed" or covered with icy "coats of iron." On the coast, the colonists had no "freinds to wellcome them, nor inns to refresh their weatherbeaten bodys," and many became sick with "colds and coughs." By spring half were dead—a "most sad and lamentable fact."

Ten years later, the colonists who would establish Massachusetts Bay Colony about fifty miles north of Plymouth, arrived in June, when strawberries were ripe for picking. While still in the process of identifying "proper places" for permanent settlement, they lived in temporary shelters—in flimsy shacks and tents or, relying on the help of Native Americans, in what the colony's first governor and historian John Winthrop described as wigwams. The poorest migrants huddled together in "miserable hovels." On December 24, Winthrop recorded in his journal that "till this time there was (for the most part) fair open weather with gentle frosts in the night. But this day the wind came NW, very strong, and some snow withal, but so cold as some had their fingers frozen and in danger to be lost." Days before, seven colonists were shipwrecked on their way from Boston to Plymouth and stranded overnight on an icy beach. Some of them had gotten "their legs frozen into the ice so as they were forced to be cut out." They lit a fire but the cold was too strong for the English and, despite the help of Wampanoags from Plymouth, the whole company was weakened. Four colonists died, their flesh "mortified by frost." In the winter of 1630–1631, over 200 immigrants lost limbs to frostbite or died from illness and exposure; another 200 left the region. In his *History of the Colony of Massachuset's Bay* (1764), Thomas Hutchinson cited a woman who visited southern New England in the mid-seventeenth century. She decided that any accounts of mild weather in the region were either unreliable or written "surely in strawberry time." During the rest of the year, she saw a relentlessly harsh environment: "the air of the country is sharp, the rocks many, the trees innumerable, the grass little, the winter cold, the summer hot, the gnats in summer biting, the wolves at midnight howling." "If you please," she retorted (alluding to Morton's account), "you may call it a Canaan."[4]

Such were the widely contrasting early descriptions of northeastern North America's seasonal temperatures, weather, landscapes, and resources. These contradictions were not merely a reflection of timing. Although Brereton and Smith came to the region during the spring, Morton and Wood stayed long enough to experience all the seasons, including the winters, which impressed many other nonnative observers as significantly longer, colder, and snowier than winters in the British Isles or parts of continental Europe.

Yet these inconsistencies are not evident in historical maps. Before the nineteenth century, the only graphic representations of global or regional climates essentially reproduced classical models based on latitude, centered on and derived from the self-proclaimed temperate empires of the eastern Mediterranean. Between early modern iterations of ancient zonal cartography and Alexander von Humboldt's isothermal maps first published in his 1807 *Essai sur la géographie des plantes*, there are no maps that allow for a visual analysis of the historical cartography of climate. As a result, this chapter explores changing ideas about climatic geography by considering the many competing textual descriptions of the location and extent of the temperate zone. It focuses on the ways in which a variety of writers imagined the Northern Hemisphere's climate in general and the northeastern North American climate in particular.[5]

Assessments of the climate in settler colonies like those that the English and French worked to establish in the Northeast tended to be divergent for a number of interrelated ideological and methodological reasons. Published descriptions of American climates were often passages in promotional literature aimed at royal patrons, investors, merchants, and immigrants. They essentially offered an environmental rationale for, or argument against, settler colonialism and land speculation in particular areas. Predictably, these pamphlets centered the temperate zone in whichever region—from Newfoundland to the Rio de la Plata—that they happened to be endorsing and, at the same time, marginalized competitors. Some writers absolutely excluded the Northeast or parts of it from the temperate zone. Other writers tried to diminish the reported differences between local and ideal weather

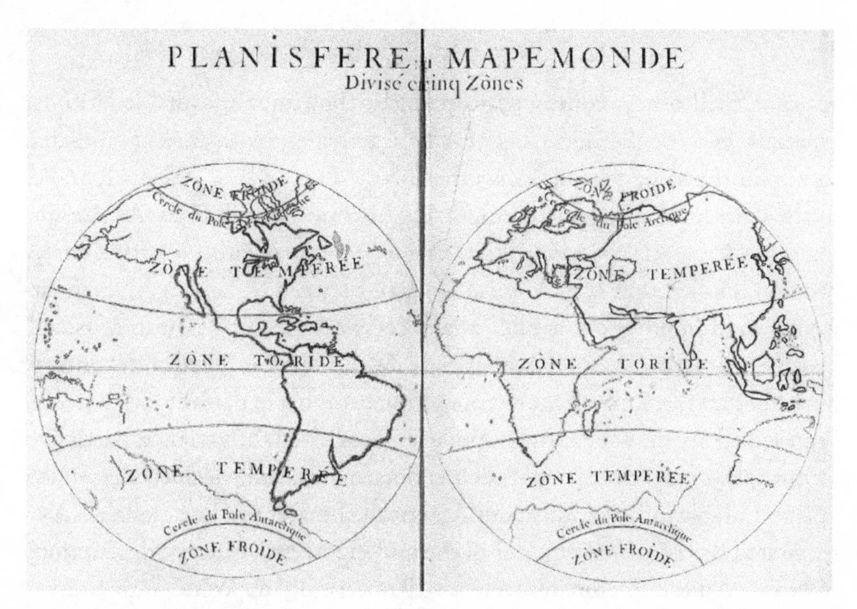

FIGURE 1.1 A seventeenth-century French map based on the classical division of the world according to five climate zones. From abbé de Dangeau, "Planisfere ou Mapemonde divisee en cinq zones" (Paris, 1693). Courtesy of David Rumsey Historical Map Collection.

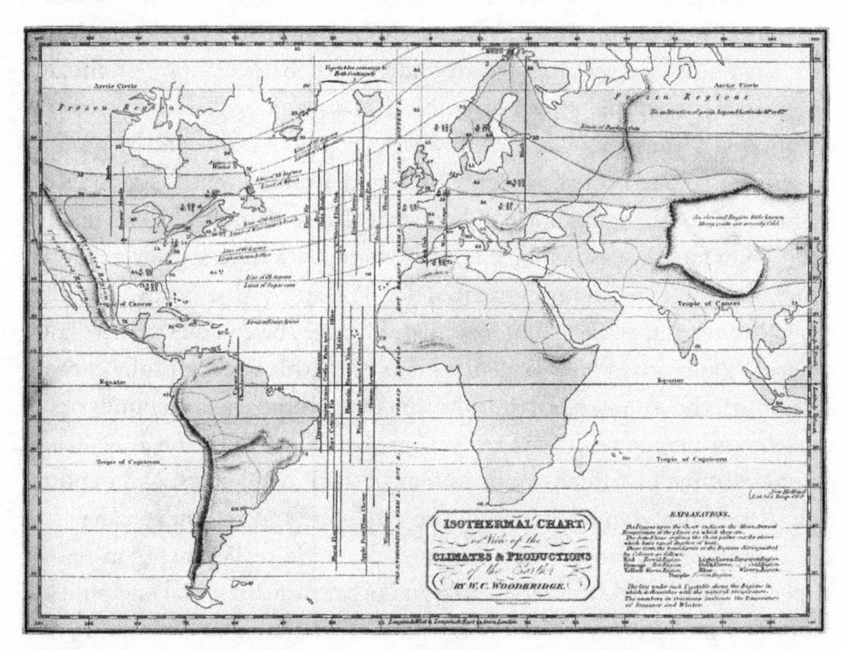

FIGURE 1.2 An 1828 map based on Alexander von Humboldt's isotherms. The inset explains: "The Figures upon the Chart indicate the mean Annual temperature of the places. . . . The dotted lines crossing the Chart point out to places which have equal degree of heat. These form the boundaries of regions distinguished by color. . . . The numbers in fractions indicate the temperature in summer and Winter." From William C. Woodbridge, "Isothermal Chart Or View Of the Climates & Productions of The Earth" (London: Whittaker, 1828). Courtesy of David Rumsey Historical Map Collection.

patterns. Still others continued to point to the region's favorable latitude, especially because the alignment of American regions with ancient cities like Rome (in the same latitude as southern New England) or Athens (parallel to the Chesapeake) suggested tantalizing historical possibilities. As a result, many never entirely abandoned the ancient equation of latitude with climate. Morton implied that New English Canaan between 40° and 45° North latitude would improve on its biblical antecedent since the "Canaan of Israel" had been situated at 30° North latitude. Well into the nineteenth century, numerous texts compared the regional climate to that of parts of Italy, Greece, France, and Spain. When contemporaries recognized the variance, partiality, or outright tendentiousness of such endorsements, they decided that much of the available information about American climates was questionable. As a frustrated British observer put it in the mid-eighteenth century, descriptions of North American climates were so "diametrically opposite"—"one says it is healthy, the other, that it is the grave of the human species"—that it was impossible to know which accounts to believe.[6]

Ostensibly more sincere efforts to characterize local climates also ran into problems. Another source of uncertainty stemmed from the instability of political borders in North America throughout this period, which further complicated early attempts to understand the geography of climate. The massive territorial sweep of political maps represented claims to sovereignty over contiguous lands. However, in practice and on the ground, imperial or colonial rule was highly contested, unstable, and geographically fragmented.[7] Although all naturalists struggled to delimit phenomena like wind, snowfall, or seasonal bird migrations that transgressed sovereign territory, most writers simply chose to describe features that occurred within town, county, imperial, or national lines. Because the colonies or states that eventually comprised New England and Nova Scotia were not fully consolidated until the second quarter of the nineteenth century, the boundaries of the relevant unit of analysis were continually changing. The ongoing demographic decline of Indians from military assault, epidemics, and deportation, the elimination of French North America, and the independence of the United States caused only the most dramatic realignments in native, imperial, or national frontiers. These and other shifts in social and political geography did not directly affect the physical climate, but early naturalists were well aware that the mutability of borderlines redrew the limits of whatever area they were studying.[8]

And yet discrepancies between conflicting knowledge of actual climatic conditions and the idealized cartographical image of climate zones persisted.

Despite centuries of mounting if equivocal evidence to the contrary, many writers on both sides of the Atlantic world represented the Northern and Southern Hemispheres' temperate zones much as the Ancients had: as large swaths of habitable environments, which now extended into the Western Hemisphere. John Oldmixon's 1708 *The British Empire in America*, for example, was a translation of a geographical work that claimed to be a "new and accurate description of the Earth in all its empires, kingdoms and states," but it presented the very old picture of climate zones:

> The Ancient Geographers divided the Terrestrial Globe into Zones, Climates, and Parallels. For as the Heavens are divided into five parts, by the four lesser Circles, viz. the Tropick of Cancer, the Tropick of Capricorn, Arctick Circle, and the antarctick, so they divided the Earth by these Circles, which lie under those of the Heavens, and exactly answer to them, into five Zones. Zones in this case, are no other than Spaces or parts of the Earth, which have different names given them according to the different temper of the Air which one breaths in them; and there are five of them, viz. one Torrid, two Temperate, whereof one is towards the North, the other towards the South and two Frigid in like Position.

To determine the "temper of the Air" in the Northeast, one could rely either on highly generalized, antiquated models of global climates like Oldmixon's or on specific but contradictory descriptions of the region.[9]

By the late eighteenth century, prolonged debates about the local climate drew attention to problems with the traditional cartography of the temperate zone. As local naturalists confronted increasing evidence of latitude's inadequacy for understanding the wide range of variations within the temperate zone, they adapted ancient concepts, arguing that the regional climate's peculiarity could be more fully understood through firsthand observation. They joined others who argued that American natural history could only be perfected by the "labours of many people" contributing detailed reports about "different regions." "To be a true Geographer," the historian Jeremy Belknap told author of *American Geography* Jedediah Morse "it is necessary like Holland to be a Traveler & a Surveyor. To depend on distant & accidental information is not safe & there is a material difference between describing a place that we have seen & one that we have not seen." "The American climate is so different in different parts of the same state, and often in the same latitude," explained Vermont naturalist Samuel Williams, "that it cannot be well understood, but by viewing it in its variations through the different parts of the northern continent."[10]

This localized, empirical approach to studying climate eventually disintegrated the ancient image of the temperate zone as a solid climatic belt girdling the Eastern and Western Hemispheres. It became implausible to picture climate zones as consolidated units, like a series of contiguous stripes encircling the Earth. In place of the static zonal model, early American naturalists' conceptions of temperate geography approached the modern notion of a fragmented and dynamic spatial distribution of regional climates—one that, if visualized, might resemble a military camouflage pattern or a scatter plot rather than a linear spectrum. Instead of a uniform loop, they imagined the temperate zone as a constellation of irregular spaces. By trying to demonstrate that the Northeast was one of those spaces, local elites helped redefine temperateness and, in the process, advanced the notion of a discontinuous geography of regional climates subject both to intellectual debate and to historical change.

A Different Latitude

In antiquity, geographical writers placed the imperial civilizations of the Mediterranean at the center of the habitable world, an ideal reflected in its name—middle of the Earth. Most early modern European maps were based on this ancient orientation, especially the second-century Greek geographer Ptolemy's division of the globe into imaginary, perpendicular lines of longitude and latitude that curved the Earth to form a lattice or, when rendered in two dimensions, a grid. The parallels of latitude defined what the Greeks called *klima*, or climate zones, so much so that geographers used the terms interchangeably: Climate was equivalent to the 180 latitudes running through Europe, Asia, and Africa, which diverged 90° north and south from the equator. Geographers further grouped these climatic belts into anthropocentric zones ranging from the uninhabitable to the hospitable. In the so-called frigid and torrid zones—two gelid regions surrounding the North and South Poles and a very hot band at the equator—people could barely survive or even visit except at great risk to their health. The only inhabitable climate, also known as the *ecumene*, supposedly lay between these extremes.[11]

Ptolemaic maps and globes thus represented the world's climatic regions with a simple spatial geometry: a striped circle or sphere, images that suggested something like a sliced and reassembled hard-boiled egg in which each slice of climate was entirely uniform. Because early modern geographers, colonizers, and migrants believed that climatic conditions across the global polar, equatorial, and intermediate zones would be more or less equivalent, after the fifteenth century, they applied this egg-slicer logic to the New World by tracing lines of latitude westward. By this reckoning, the annual climate on the coasts between the Northumberland Strait and Long Island Sound should be the same as the temperate climate on the coast of Provence. Farther to the north, the climate would resemble that of northwestern Europe. In the late sixteenth century English explorer Martin Frobisher explained that his countrymen should focus on colonizing Greenland, Newfoundland, and Labrador, where it was as temperate as in the British Isles, and avoid exploring equatorial regions in the Americas, where the tropical climate was dangerously hot. Besides discoveries of gold, silver, and a northern sea route to Asia, Frobisher foresaw that they could establish a trade in wool to supply the domestic needs of indigenous people. It also happened that these northern lands were supposedly "not alredie possessed or subdued by or to the use of anie Christian Prince in Europe," so the English would face less competition there. "By God's providence," the northern territories of the "newefounde worlde" had been reserved for the English empire.[12]

However, as French explorer Jacques Cartier had noted in the 1530s, although ancient philosophers were undoubtedly wise, their idea of the five global climate zones had never been put to the "test of actual experience."[13] Renaissance polymath Jean Bodin's influential *Method for the Easy Comprehension of History* (1566), which linked natural and political history through the abiding influence of climate, integrated a variety of explorers' American experiences to argue that geographical coordinates proved inadequate to account for all of the world's climate zones. Encounters with indigenous societies from Patagonia to Hudson Bay showed Europeans' preconceptions to have been overconfident and naive. Although in theory there were two temperate zones, ancient and medieval geographers focused mainly on the Northern Hemisphere's *ecumene*, about which they knew more and with more certainty than they knew about the temperate climates of the Southern Hemisphere. Reports of Iberian and French exploration and conquest in the humid, lush tropical forests of South America challenged the notion that the excessively hot climate of the torrid zone was so unfavorable to most forms of life that even a transit between the Northern and Southern Hemispheres could be lethal. Frenchmen, for example, were surprised and delighted by the nature of early seventeenth-century Maragnan in northeastern Brazil, which they had expected would be a scorching hot, lifeless desert. Instead, one Capuchin missionary believed that he had never been in "a place more temperate and delicious than that country." Fishermen and voyagers searching for a northwest passage to Asia were perhaps less surprised to find local residents in latitudes corresponding to Ireland in North America but also less delighted by what they found: frosts unlike any they remembered from home. Forays into the waters and along the coasts of the Labrador Sea, Hudson Bay, and the Gulf of Saint Lawrence struck the English with discomforting visions of ice mountain islands, even in July. Nevertheless, the "Eskimaux" managed to inhabit these frigid lands throughout the year.[14]

Contrary to long-standing assumptions, neither the equatorial nor subarctic regions of the Americas were too extreme for human life, at least for some societies. Ancient geographers had "fancied" the tropics "to be Uninhabitable by reason of the Excessive Heat," wrote royal hydrographer and globe maker Joseph Moxon in 1701, "but we have now discovered their mistakes, and find not only the Cold Countries somewhat near the Pole, but the warmer Regions under the Equator to be Plentifully Inhabited." Over time, the divergence of American climates from European expectations radically changed conventional models of the *ecumene* in two ways: first, by extending the map of the inhabited world to the Americas, including areas in the tropics and Southern Hemisphere and, second, by discrediting the idea that climate zones were absolutely parallel.[15]

However, transcending the limited geography of the classical *ecumene* did not erase the perceptible differences among polar, temperate, and tropical climates, including subtle differences within these zones. Frobisher's experiences convinced him that the disadvantages of traveling in tropical heat were hardly as numerous or extreme "as those of the colde proceding in the Northe," which lasted all day and night (so it was little wonder that locals continued to prefer fur over the knitted clothing the English offered them). Cabot, Frobisher, Gilbert, Hudson, and other early adventurers were either too reluctant to learn how or died in the process of trying to live in the inclement and enduringly cold weather of far northern regions.[16] In territories south of the Gulf of Saint Lawrence explorers and settlers were more perplexed by annual weather conditions. In September, 1602, Bartholomew Gosnold wrote to his father from the Gulf of Maine to describe:

> that place where we were most resident, it is in the Latitude of 41 degrees, and one third part; which albeit it be so much to the Southward, yet is it more cold then those parts of Europe, which are scituated under the same paralell: but one thing is worth noting, that

> notwithstanding the place is not so much subject to cold as England is, yet did we finde the Spring to be later there, then it is with us here, by almost a moneth.

Gosnold wondered whether the long winter "hapned accidentally this last Spring to be so, or whether it be so of course." Without a historical comparison, he was "not very certaine" but decided that "the latter seemes most likely."[17]

Others substantiated his guess about the severity and persistence of cold weather in the Northeast, particularly during the worst winters of the Little Ice Age in the seventeenth and eighteenth centuries when colonists routinely suffered painfully low temperatures, deep freezes, arctic winds, heavy snowfall, slick paths, and icy waters. Worse still, these conditions were long lasting: In bad years, wintry conditions could make their appearance as early as September and persist into late May, or retreat in March and then unexpectedly return with full force in April. Following the collapse of the early seventeenth-century English colony at Sagadahoc, which was "begun and ended in one yeere" because of the "extreme frozen Winter," the Maine coast was "esteemed as a cold, barren, mountainous, rocky Desart." The Plymouth colonists relied on the charity of Wampanoag farmers in the early 1620s because of the newcomers' "negligence" but also their unpreparedness for the "extremtie of weather, both of wind, snow and frost" that occurred in May and June during those years. "They that know the winters of that country," Bradford wrote about Cape Cod, "know them to be sharp and violent, and subject to cruel and fierce storms."[18] In 1749, winter's onset prompted one Massachusetts clergyman to pray for a mild season. The air was already "pritty Cold" in early November; by the end of the month he wrote in his diary that "the Season looks at present very terrible. Tight shut w'th snow and some Days very Cold. It fills many (I suppose) w'th deep Concern." That year hard frosts continued through spring, cold enough "to freeze my Ink on the Table," wrote another minister once his ink had thawed. There was "so much cold weather in New-England; so many snows & river-freezings," wrote a naturalist to his family in Rhode Island in 1799, that he seriously began to consider relocating to the Ohio Valley "where undoubtedly, the seasons are more temperate."[19]

Compared with southern New England winters, Acadia/Nova Scotia's were judged by an official to be "much colder." According to several sources, all the trees from Montreal to Halifax were "covered with icicles, which are even dangerous to those that stand under them; there is no more stirring out of doors, without being wrapped up in fur, and in spite of this precaution, not a winter passes without loss of limbs by the benumbing cold." Added to the hazards of "long and intense colds" were thick fog and heavy snowfall, which caused whiteouts so overwhelming that "the eye is pained with an universal whiteness" that blurred "any difference between land and water." The Scottish minister Andrew Brown believed that, because of its long winters, Nova Scotia experienced only three seasons. "The rigour of the winter, and the shortness of the summer" made spring nonexistent: "The season of Spring, the most beautiful of the year is scarcely marked and was never enjoyed in this part of the new world." As a result, wheat and grapes, the key ingredients for communion, were never locally grown. Although "in the parallel of the richest grape of Europe, but the vines which were planted in Acadia in the same latitude from which they were taken perished in the frosts of the first Winter."[20]

Already by the late sixteenth century, explorers had written sufficiently discouraging reports of winter conditions in New England and Nova Scotia that some considered them as part of "those North and colde regions" stretching to the Arctic Circle. But compared with the far north, which was clearly in the frigid zone, the Northeast's annual climate was more difficult to categorize absolutely in terms of zonal geography. If winter and spring conditions made the region feel as though it was situated close to the Arctic, during the short summer growing season it could seem subtropical. A mariner's guidebook tried to resolve this problem by proposing that American regions north of 42° North latitude—that is, north of Cape Cod—were not Mediterranean but instead part of a hitherto unknown "temperate Arctick zone." Few writers picked up on this reimagined geography, dwelling instead on the region's seasonal extremes. In his 1720 *History of New England*, which included Nova Scotia, Daniel Neal wrote that, during summer, "the people are very much troubled by mosketoes or little flies like gnats, bugs and other insects that delight in Heat." Neal's assessment was echoed in 1774 in a private letter from a newcomer to Nova Scotia to his brother in Scotland: The province's seasonal conditions usually meant being "tormented all the summer with mus keetoes and almost frozen to dead in the winter." According to Samuel Williams's daily thermometer readings in Cambridge, seasonal temperatures ranged from −14° to highs of 99° Fahrenheit. Charles Blagden, a physician and member of the Royal Society, was "struck" by these "extremes" when he was stationed with British troops in Newport, Rhode Island, during the Revolutionary War. "What an abominable climate!" he exclaimed in a letter to Sir Joseph Banks in 1778, "particularly in the winter."[21]

As such unfavorable accounts continued to circulate, they slowly eroded belief in the regional climate's equivalence with Old World climates in the same latitude. Like many chronicles, William Hubbard's *General History of New England* (1680) included a chapter on "the Temperature of the Air and Nature of the Climate" and began with a common refrain about the region's intermediate position according to latitude:

> the climate of New England lyes in the middle between the frigid and torrid zones, the extrems on either hand; and therefore may bee suposed to bee in the most desirous place of a temperate ayre, for the advantage both of wholesome and delightfull living, falling into the same latitude with Italy and France: some provinces in both which countrys in former times being taken for the most desirable in the whole universe.

However, "longe experience" proved this supposition to be false. "By reason of some occult and secret accident," Hubbard wrote, the summers and winters "annoyed" its inhabitants by partaking "a little too much of the two extrems of heat and cold, proper to the two opposite regions on either hand." In sum, wrote Neal, the region's climate was "not so temperate as ours in England, their Summers being shorter and hotter and their Winters longer and colder; nor is it so mild and regular as those Parts of Italy and France, that lie in the same parallel in Europe." Yale College president Timothy Dwight extended the transatlantic comparison to other parts of continental Europe:

> The great distinction between the Climate of New-England, and that of European Countries lying in the same latitudes, i.e. of the countries lying between Oporto, Barcelona, Naples, and Constantinople, on the South; and Buda, Munich, Brisac, and Port L'Orient on the North, is the peculiar coldness of its winters. The heat of our summers may be somewhat greater, than that of these European Countries. The cold, at times, is unquestionably much greater.[22]

If for Hubbard, Dwight, and Neal these local conditions were merely an unpleasant feature of life in the region, for critics of northern colonization they presented serious disadvantages. The "singular and peculiar" climate throughout coastal North America translated into "a difference of at least fourteen or fifteen degrees of latitude between the respective climates in these two continents; it being so much colder there than" in Britain, wrote John

The Seamans Secrets.

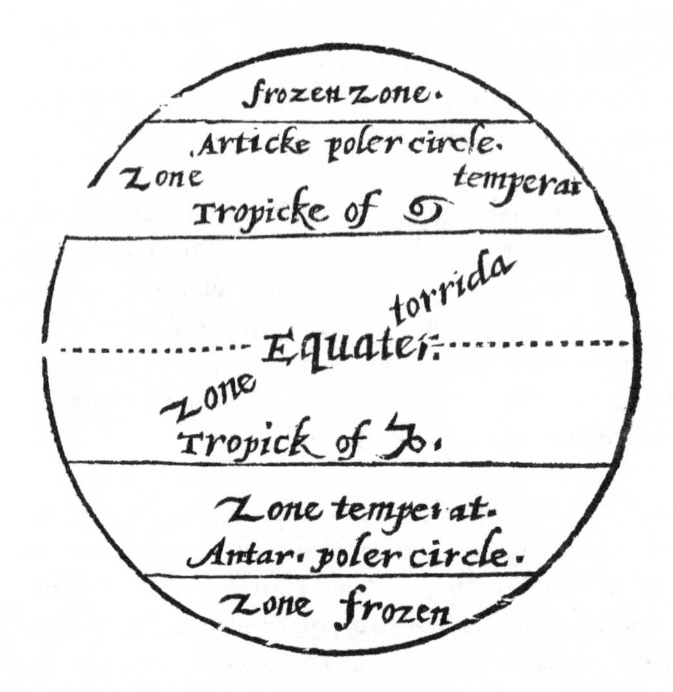

What is a Climate?

A Climate is the ſpace or difference vpon the vpper face of the earth, included betwéen two parallels, wherein the day is ſenſibly lengthened or ſhortened halfe an houre, for as you trauaile from the Equator toward the Artick Pole, the Sunne hauing North declination, the dayes doe grow longer and longer vntill at laſt the Sunne not ſetting vnder the Horizon, you ſhall haue continuall day, and euery ſpace or diſtance that altereth the day halfe an houre, is called a Climate theſe : Climates take their names from ſuch famous places as are within the ſaid Climates, of which there are nine, as by their diſtinctions may appeare.

1. The firſt paſſing through Meroe, beginneth in the latitude of 12. deg. 45. m. and endeth in 12. d. 30. m. whoſe breadth is 7. d. 45. m.

C 2 2. The

FIGURE 1.3 This book posited a "temperate arctic zone." From John Davis, *The Seamans Secrets, Divided into Two Parts* (London, 1599), n.p. [part 2] ("temperate arctic zone"). Courtesy of the John Carter Brown Library at Brown University.

Mitchell in his 1767 treatise *The Present State of Britain and North America*. Mitchell, the Virginia-born physician and naturalist best known for his 1755 map produced for the Board of Trade, explained to his patrons that it was "for this reason that the nation is disappointed, and everyone is so much deceived about North-America." The deception was acute for the northern colonies in particular, where the climate "is much worse than is generally apprehended." The combination of nearly subarctic winters and brief warm spells, Mitchell believed, meant that profitable agricultural development was impossible there. "Were it not for the fisheries, none could live in these northern parts of America."[23]

Numerous authors repeated verbatim Mitchell's argument about the unfortunate mismatch between latitude and climate in northeastern North America, as well as its economic implications. Each chapter of *American Husbandry* (1775), one of the most widely circulated comparative surveys of the climate and natural resources on the continent, began with the geographical coordinates of individual colonies. But in the first chapter its anonymous author, who liberally excerpted from Mitchell's *Present State*, warned readers that "to judge of the climate of Nova Scotia by the latitude would lead any person into egregious mistakes":

> Between 44° and 50° of latitude in Europe we find some of the finest and most pleasant countries in this quarter of the world; but in Nova Scotia the case is very different. The winter lasts seven months . . . the deepest rivers are frozen over in one night, so as to bear loaded wagons; snow lies in some places ten feet deep, and upon level tracts it has been known to be six feet deep. . . . When this severe winter goes, at once comes a summer (for they have no spring) of a heat greater than is ever felt in England. . . . But what is almost as bad as the extremes of heat and cold, are the perpetual fogs, which render the country equally unwholesome and unpleasant.

In this author's judgment, northern New England was hardly much better because it was "under the fatal influence of that freezing climate, which is bad enough in the south parts of New England but here approaches to the severity of Nova Scotia." He predicted that "this is, and ever will be, the consequence of colonizing in such northern latitudes, where agriculture must ever be carried on with feebleness; where the climate is to the last degree rigorous; and where every spot is inhospitable and frigid. To plant colonies in such situations, is acting contrary to every rational idea of colonization."[24]

The Northern Temperate Zone

Such wholesale condemnations of the regional climate implied that colonists had settled outside or at the far edge of the temperate zone, a particularly worrying possibility in part because of Europeans' enduring attraction to the theory of geographical determinism. Even as they began to lose faith in the universal rule of latitude, they remained attached to this axiom of classical geography: that environmental conditions reciprocally influenced inhabitants' fitness, temperament, and civilization (or lack thereof). If natural disadvantages like a severe climate—regardless of latitude—restricted or depressed growth, favorable material preconditions accelerated economic, political, and cultural development. Early explorers routinely examined the physical and cultural effects of local climates on the bodies, behavior, and village life of Native Americans through this lens. In the early seventeenth century, before imported contagious diseases, war, and colonial expansion ravaged so many of the Indian polities of southern New England, Brereton, Gosnold, and Verrazzano all reported hearty, long-lived people on Cape Cod and neighboring islands. Early English visitors were impressed by the strength and vigor, if not always the economic or cultural achievements, of Native Americans. For optimists, Indians' apparent physical fitness and their "gardens and cornefields" meant that New England had "as healthfull a Climate as any can be," where the English found "health & strength all the while we remained there."[25]

But the many reports of hard winters suggested a more troubling line of thought. The heartiness of local populations indicated that the region's formidable nature, like that of any northern climate, strengthened but also probably stupefied people. Andrew Brown, who described "the extreme rigour of the winter" in Nova Scotia, was intrigued by *Histoire de la Nouvelle-France*, Marc Lescarbot's 1609 account of French colonists who were "more delighted with the natives than the soil or its productions" because the Mi'kmaqs' "features were not so harsh and their temper was not so ferocious as those of their neighbours." To resolve this paradox, the colonists decided that "tribes inhabiting Akadie" must have migrated northward, probably from Narragansett Bay—a place with "a milder climate and a more improved state of society than that in which they were found." Overall, however, very few European commentators perceived, as Brereton did, that these "goodly people" were both physically "perfect" and "very wittie." More typically, they presumed that the Northeast's indigenous peoples were insensitive to all varieties of pain, including their lack of suffering from "inclemencies of the weather." Impassivity was not peculiar to Native Americans. Instead, the degree of a people's sensitivity was proportional to the prevailing temperatures in different regions. In his synoptic *Spirit of the Laws* Montesquieu wrote that "in cold countries they have very little sensibility for pleasure; in temperate countries they have more; in warm countries their sensibility is exquisite." A sufficient example of the mind-numbing effects of excessively cold air, Montesquieu joked, was that "you must slay a Muscovite alive to make him feel."[26]

What is remarkable is that France and England had once also been vulnerable to such formulations. From the perspective of the classical world, northwestern Europe was located on the cold, distant fringes of high civilization. The mild climate of the Mediterranean produced natural leaders. "The intermediate situation of the Greeks," explained Aristotle, "seems to have happily blended in their character the virtues of courage and prudence, and to have formed them for thinking calmly, yet feeling strongly." By contrast, unsophisticated, politically weaker societies lived at or beyond the borderlands of ancient empires. North and south of the Mediterranean, intense climates allowed for only a bare subsistence and so retarded progress of any kind. Numerous classical writers imagined the northern countries as oppressively cold, windy, damp, and barren wastelands, virtually uninhabitable except by fierce, dimwitted nomads like Gauls and Britons (or "Scythians") and cold-hardened plants and animals, that were each uniquely suited to the rigors of this environment. Hippocrates stated that, because it was "very difficult to live" in lands "oppressed" with constant cold, northerners were generally the "least fruitful" of all people. According to Aristotle, this climatic tyranny meant that "the inhabitants of Europe and of most cold countries abound in strength and courage, but their intellectual powers are feeble or defective. They enjoy liberty but are unacquainted with good polity; and though capable of maintaining their independence, are unworthy of aspiring to empire."[27]

Through the Renaissance, allusions to the northern countries could still evoke these images of weakness and incivility, which derived from the classical imperial division of the Old World. The enduring negative connotations of the North combined with England's and France's historical sense of distance from the center of empire provoked anxieties about their peripheral place in the early modern world. It also encouraged a fervent investment in the temperate ideal. As capital and long-distance trade networks increasingly shifted westward from the Mediterranean to the Atlantic world, intellectuals worked to revise the classical geographical idea that civilization was centered in Athens, Rome, or Jerusalem and to reverse, attenuate, or rotate the connotations of north and south. Coincident with western European states' earliest attempts to colonize the Americas, writers became interested in revising old stereotypes about the North. By the late sixteenth century, they began to explicitly manipulate geographical ideas about the extent of the five climate zones.[28]

Retaining the ancient principle that geographical and cultural conditions determined one another, English and French writers cannily included their locales within the temperate zone's range. The frigid north and barbarian lands lay elsewhere, they claimed. England and France were located well within the "virtuous mean between two vicious extremes," in climate zones healthy for body and mind. Such conceptual manipulation of climatic geography was plausible partly because of the wide range of seasonal temperatures that were experienced in temperate environments. The temperate zone was never too hot or too cold, at least not for any considerable span of time; what qualified as just right, however, was highly subjective. If English spring and summer were sometimes somewhat cooler compared with continental European seasons, England's winter was certainly warmer than Russia's. The English claimed that their climate's slight chill exerted positive physical and moral influences on the nation. Its cool temperate climate was bracing enough to produce not only strong constitutions but also self-restraint and rationality. In effect, they discursively extended the cultural and climatic cartography of the temperate zone to include themselves.[29]

By the eighteenth century, French and British intellectuals took for granted this expansive, revisionist idea of temperate geography and their place within it. Edward Gibbon retroactively included Britain in his picture of the ancient world's temperate zone. In *History of the Decline and Fall of the Roman Empire*, intended as an object lesson for his countrymen, he situated the ancient empire "in the finest part of the Temperate Zone, between the twenty-fourth and fifty-sixth degrees of northern latitude; and that it was supposed to contain above sixteen hundred thousand square miles, for the most part of fertile and well-cultivated land." This geographical description was sufficient to give his reader a "general idea" of his subject—a huge empire that spanned over 30 degrees of latitude, all of them in the most temperate climates of the world. Montesquieu included northern and western Europe in the temperate zone as if it were a kind of natural law. To demonstrate the relationship between political liberty and the "nature of the climate," he stated repeatedly that in the Old World truly free nations existed only in Europe, a region that stretched from Spain and Italy to Norway and Sweden. This "very extensive" temperate zone implicitly included France and Britain. Outside of Europe, brave "little free nations" like parts of northern China and Korea sometimes emerged in colder climates closer to "the Poles," but never near the Equator; in general, however, there was "no temperate zone" in Asia or Africa.[30]

Montesquieu was reticent about the New World's climatic regions, but other Europeans began to integrate evidence produced by decades of colonization and natural history collecting to test how climatic determinism might explain perceived differences in nature and culture on either side of the Atlantic Ocean. George-Louis Leclerc, comte de Buffon's secular, historical "theory of the earth" involved the long-term effects of both geological and anthropogenic climate change, migration, and topographic isolation. Buffon boldly corrected ancient geography by dividing the world into four parts. This quartet was a neat analogy, Buffon noted, to the seasons of temperate climates. Instead of "alternate bands or belts, parallel to the Equator," he proposed two oceans surrounding two terrestrial belts of the Old and New Worlds, each quarter measured from the North to South Poles. Although the terrestrial belts were "nearly" equal in length, he found that "the Old Continent extends farther north of the equator than south; but the New, farther south than north." As a result, using latitude to compare these belts' environments was wrong. At the same time, Buffon retained the deterministic prejudice against extreme climates. He speculated that the world's oldest, most developed, and largest plants and animals originated in the North Pole when the global climate was warmer; as it cooled, these species moved southward through the older landmass to seek the warmest climates of Africa and Asia. Because of its position with respect to the equator, the New World and its native biota were altogether younger, less mature, and smaller. While Buffon recast the global geography of climate, he reaffirmed long-standing certitudes about its determining influence on the distribution, diversity, and vigor of plants and animals, including the influence of climate on people's character. Because North America was cooler than Europe and hotter than Africa at the same latitudes, he concluded that over the long timescale of the Earth's history, as Old World species developed toward "perfection" in moderate climates, their New World analogs had degenerated in the Western Hemisphere's relatively harsher conditions.[31]

Cornelius de Pauw, the Abbé Raynal, Oliver Goldsmith, and William Robertson were among the more widely read and cited European writers who further popularized Buffon's polarizing theory about the New World, a theory with dismaying implications for North America, particularly the viability of settler colonialism in the Northeast. According to Goldsmith's *The History of the Earth*, in a cold hemisphere, the "cold frozen regions of the north" stunted all living things; with the exception of bears, native species in the northern latitudes of North America were especially small by nature and immigrants quickly diminished in size. Even bees and spiders were "not half so large as those in the temperate zone." In "the northern provinces of America," wrote Robertson in *The History of America*, "nature continues to wear the same uncultivated aspect, and in proportion as the rigour of the climate increases, appears more desolate and horrid."[32]

Writers in the Northeast balked at such transparently chauvinistic ideas. Responding to Robertson's assertions about the region by reversing them, Samuel Williams argued that Vermont had "a power, an energy, a vigor in the vegetable life which nature has never exceeded on the other side of the Atlantic." Rather than pathetic creatures, he claimed "the climate of America seems to be peculiarly suited to the most mild, temperate, and useful animals." The "misrepresentations" of the region written "by European travelers" motivated Timothy Dwight to begin writing an epistolary narrative of his own experiences in "the Northern States." Dwight bristled at the "misrepresentations, which foreigners, either through error or design, had published of my native country." These outsiders drew a depressing if by-now-familiar "caricature" of the local environment as cold, barren, poor, and unpromising. "Silence under such aspersions is easily construed into a confession of their truth." Perhaps the winters were somewhat longer and colder in New England and New York, Dwight conceded, but this seasonal variation was typical of other areas "in the Northern temperate zone, perhaps throughout the world." As the English had done in earlier centuries with regard to the British Isles, learned settlers like Dwight insisted that "the weather of this country" was not aberrant but instead part of a newly reconceived northern geography of temperate climates.[33]

Natural and Political Histories of the Northeast

As writers like Dwight became entangled in broader eighteenth-century disputes about the New World's supposedly severe, inferior, and regressive climate, they increasingly emphasized the necessity of accurately documenting local conditions. But what were the spatial limits of the local climate in the Northeast? For his part, Dwight never specified the extent of his "Northern temperate zone," the position of the region within it, or even the precise geography of the region. While he was careful to delineate the exact boundaries of the New England states, his work also encompassed neighboring states (published posthumously, it was titled *Travels in New England and*

New York). Like others of its kind, Dwight's work indicated the geographical ambiguity of the region's outlines, a predicament that stemmed from the constant flux of the region's political boundaries. The place that was sometimes called New England at various times encompassed all or parts of present-day Nova Scotia (including Cape Breton Island), New Brunswick, Prince Edward Island, Vermont, New Hampshire, Connecticut, Rhode Island, Massachusetts, Maine, and the many smaller islands near the region's Atlantic coast; less frequently, it also included Newfoundland, Quebec, and New York. Because Native American, Dutch, French, British, and American claims to the region continually expanded and contracted over this period, there was no general agreement on the boundaries of the region's climate. Officials, geographers, surveyors and naturalists who wrote about environmental features often referred to the territory they described as "these northern latitudes," "northern climates," "northern states," "northern parts," "northern colonies," "the more Northern Provinces," "Eastern Country," or just "this country" without a consistent or specific definition of its constituent parts and outer bounds.[34]

It should be unsurprising that Europeans seldom referred to the region's political or natural geography according to Native American concepts. Because so much of the regional economy was oriented to maritime activities, Europeans had a more accurate understanding of the well-traveled coasts at the edges of the cod and whale fisheries—especially those areas that native groups inhabited throughout the year in the early colonial period—and of port towns than of the interior or uplands. Where native villages and networks dominated the landscape, colonial travelers, traders, and settlers often failed to recognize distinct groups or confederacies, let alone their rivalries with one another. By the middle of the seventeenth century, impressionistic colonial reports downplayed this ignorance of Indian country by emphasizing the dwindling numbers or asserting the complete absence of indigenous populations in the region. A rare exception that proved the rule was Nova Scotia naturalist Titus Smith, Jr.'s early-nineteenth-century description of a Mi'kmaq method for "meteorological predictions" based on observations of animal migrations, which he expressed in the past tense because he believed that most Indians had either left the province or "their number must be decreasing." Settlers' sketchy grasp of indigenous presence or Native communities' conceptions of space reflected the ways in which brief encounters, conquest, war, epidemic disease, and trade dominated Native–European interactions, rapidly and often violently reshaping the Northeast's political, cultural, and ecological geography.[35]

This colonial erasure was also an act of profound appropriation. Settlers retained some Algonquian place names for villages or natural features, but they were quick to rename the entire region, usually after themselves. The English briefly called it Norumbega—a pseudo-indigenous toponym with mysterious origins. However, most of the earliest English maps and charters designated the coastal territories north of the Chesapeake Bay as North Virginia. In 1616, John Smith clearly signaled his country's imperial aims when he appropriated part of this area as New England, the term James I used four years later when he gathered a group of London investors to form the Council for New England, which was responsible for a territory in America that stretched from "fourty to fourty eight degrees of Northly Latitude, and in length by all that breadth aforesaid from Sea to Sea." James I then extended this branding technique northward when he granted the peninsula that the French called Acadia to William Alexander, whose New Scotland colony lasted only from 1629 to 1632 but provided later British colonizers with the latinized name Nova Scotia.

Alexander's short-lived colonial project left unresolved the border with Acadia and New France, the first in a series of disputes over imperial sovereignty in the region, including English colonists' own conflicting ideas about

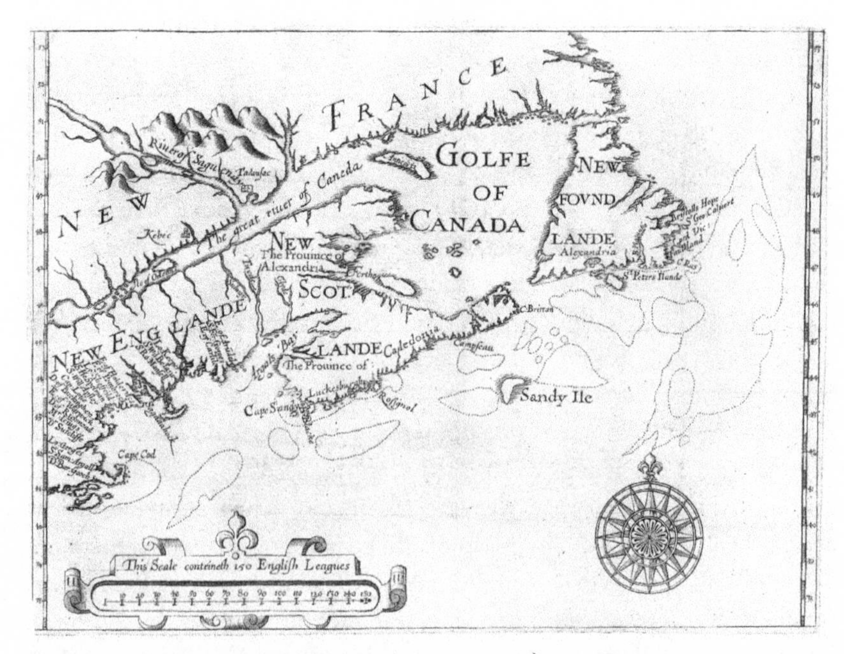

FIGURE 1.4 William Alexander, *An Encouragement to Colonies* (London, 1624) [n.p.]. Courtesy of the John Carter Brown Library at Brown University.

New England's boundaries. Complications resulted partly from the Council of New England's expansionist mandate but also from its members' ignorance of the geography of the country. Most of them had never crossed the Atlantic. They might have settled these border disputes with reference to maps based on latitude if not for England's dubious if common practice of making overlapping land grants: The original charters for the Plymouth and London branches of the Virginia Company shared claims to American land between 38° and 41° North latitude.[36] This may be why John Josselyn, who visited Boston in the 1660s, believed that New England meant every English colony from 40° to 46° "Northerly Latitude, that is from the de la Ware Bay to New-found Land"—the same vast territory that France had claimed as its own and that included Acadia and Île Royale (later British Cape Breton Island). Because Josselyn also mentioned the possibility that New England was a large island within the larger island of America and was inhabited by indigenous Tartars or Samoeds, perhaps he was an unreliable source on regional geography.[37] Nevertheless, James II imagined a region of similar scope when in 1686 he revoked the Massachusetts Bay Colony's proprietary charter and created a supercolony modeled on Spain's viceroyalties that included all British lands from the Delaware River to New France. The Dominion of New England, as he called it, lasted only until 1689. Soon thereafter, ambitious colonists in New England and New York began organizing a conquest of the Catholic French to their north. A partially successful campaign against the French turned Acadia over to the British in 1710, formalized in the 1713 Treaty of Utrecht.

Despite the treaty, surveyors charged with recording official boundary lines were reluctant to act decisively in settling rival claims. In 1720 the British commissioned French Huguenot engineer (later governor of Nova Scotia) Paul Mascarene to survey Acadia/Nova Scotia's boundaries. Mascarene described a territory that stretched: "from the Limits of the Government of Massachusetts Bay in New England or Kennebeck River, about the 44th degree North latitude to Cape des Roziers on the South side of the entrance of the River of St. Lawrens in the 49th degree of the same latitude, and . . . from ye Easternmost part of the Island of Cape Breton to the South side of the river of St. Lawrens." Although he could offer precise degrees of latitude, Mascarene cautioned that his account portrayed the colony only "according to the Notion the Brittains have of it." Because "the Boundaries having as yet not been agreed on between the Brittish and French Governments in these parts," he maintained that, "no just ones can be settled in this Description." Daniel Neal's book, published in the same year, offered the most inclusive

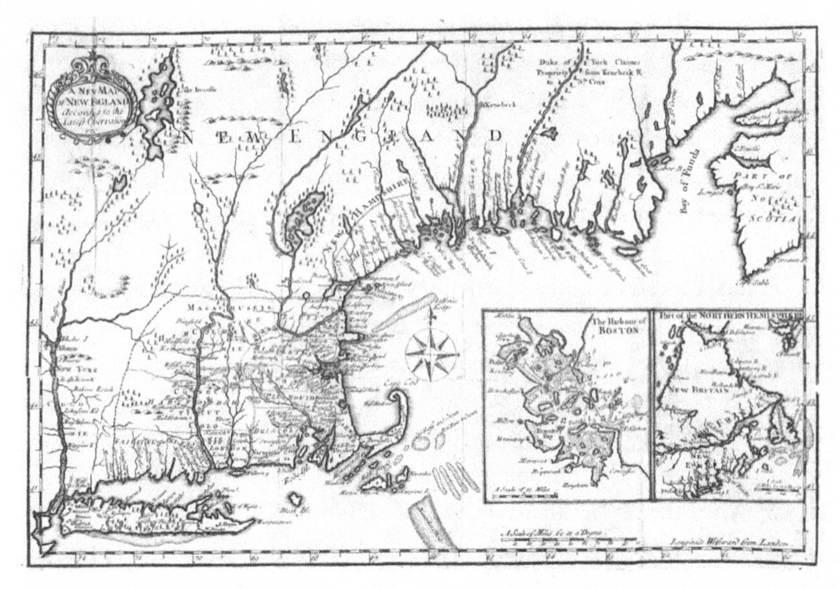

FIGURE 1.5 Microcosm and macrocosm: "A New Map of New England According to the Latest Observations," (1720). From Daniel Neal, *History of New England*, Vol. 1[n.p.]. Courtesy of the John Carter Brown Library at Brown University.

notion of the colonial Northeast in terms of competing European claims (though in terms of latitude, it did not extend as far north as Mascarene's description): "New England is that Part of the continent of America, which lyes between Degrees of 41 and 45 North Latitude. Under this Name is usually comprehended the colonies of the Massachet Bay, New Plimouth, and Connecticut; the Provinces of Main, New Hampshire, Rhode Island, Providence Plantation, and Accadia, or New Scotland."[38]

The region's outer bounds were still more dramatically enlarged after the Seven Years' War when France ceded all of its mainland North American possessions to the British. But even before that victory, British cartographers and historians deployed New England's earliest charters to lay claim to lands much farther to the region's north and west. In his influential 1755 map for the Board of Trade, John Mitchell noted the "Northern Bounds of New England by Charter Nov. 3rd 1620, extending to the South Seas" just below the section picturing Lake of the Woods in "north central North America." In the midst of the war William Burke wrote that, although it was "somewhat difficult to ascertain the bounds of the English property in North America, to the northern and Western sides," if France were ever to be ejected from the continent, "to the Northward, it should seem, that we might extend our claims quite to the pole itself." After 1783, this expansive vision was still possible, but Britain's contiguous territorial claims to the Arctic would extend northward only from its new border with the United States.[39]

During the perpetual reshuffling of sovereignty throughout the seventeenth to early nineteenth centuries, the interior and exterior borders of the Northeast grew, shrank, or changed so frequently that charters, surveys, maps, and geographical references sometimes simply situated New England or Nova Scotia somewhere in the "more northerly parts" of North America. Even after Smith's 1616 coinage became the standard name for the region, the geographically indeterminate if monolithic name North Virginia continued to be used as another shorthand. The signers of the Mayflower Compact invoked it in 1620, when they asserted that Plymouth Plantation would be "the first Colony in the northern Parts of Virginia." Sometimes images and descriptions of North Virginia included adjoining areas of New France to the northwest or the Middle Colonies to the south, or it was used in an even broader sense "as a general name for the whole Northern Continent." As late as the 1750s, Boston physician and historian William Douglass substituted North Virginia as a collective term for New England and Nova Scotia in his widely circulated geographical history. He justified this usage by citing the many maps and other texts that continued to use it. And, in 1767, John Mitchell included everyplace north of Virginia in his pessimistic description of the climate in all "the northern colonies."[40]

If Douglass exaggerated the currency of North Virginia as a place-name, eighteenth-century writers called the region by other names that emphasized northernness. References to the north could reflect economic geography, as when Arthur Young disparaged the commercially unpromising lands of "the Northern Colonies," adding the clarification "that is, north of tobacco." Such geographical lumping also reflected political geography, particularly new administrative units that the British improvised over the course of the Seven Years' War. In 1755, the Board of Trade used the Potomac River (regarded as the center of the continent) to divide the administration of colonial territory between a northern and a southern Superintendent of Indian Affairs. Perhaps this is why at the end of the war surveyors were assigned to a Northern and Southern District—divided now at Pennsylvania—for the purpose of creating comprehensive maps of these two regions that integrated former French and Spanish territories Quebec, Cape Breton, Prince Edward Island, and the Floridas into the map of the British Empire in North America. Between 1764 and 1775, the Board of Trade and the Admiralty assigned military engineers and Seven Years' War veterans Samuel Holland and Joseph Fredrick Wallet DesBarres to survey the lands and coastline of the Northern District between Rhode Island and Quebec. Largely as a result of their detailed work, particularly the publication during the late 1770s of DesBarres' coastal survey as a maritime atlas, *The Atlantic Neptune*, the term Northern District or variants of it appeared on subsequent maps like one published in 1780 depicting the territory of Vermont's contested boundaries with "part of Canada" and the surrounding states of New Hampshire, New York, Connecticut, and Pennsylvania. This broad area was labeled "the Northern Department of North America." As the circulation of these place-names implies, writers and cartographers often resorted to designating the region as lying north or northeast of someplace else—Virginia, Maryland, Pennsylvania, tobacco plantations—rather than as any more specific entity.[41]

After American independence, the northern borderlands between the United States and British North America, particularly between New England and Nova Scotia, presented continuing problems for geographers. Although Jedidiah Morse vowed that *American Geography* (1789) was "as accurate, compleat, and impartial as the present state of American Geography and History could furnish," he could not definitively describe the geography of New England. In the early national period, the region south of British North America unquestionably comprised five states—Connecticut, Massachusetts, New Hampshire, Rhode Island, and Vermont—and the District of Maine, which until 1820 remained a noncontiguous part of Massachusetts. But Morse was unsure of how to address the disputed waters off the coasts of Maine and peninsular Nova Scotia or the territory between Maine and New Brunswick. In 1784, Britain hived off parts of Nova Scotia to establish the separate colonies of Cape Breton and New Brunswick but Morse may have been unaware of the boundary created between New Brunswick and the District of Maine in the same year (in any case, the boundary was not officially settled until the 1842 Ashburton–Webster Treaty). Through the 1790s, maps continued to show New England as a whole encroaching well into this new British province. In addition, in the early 1780s there had been plans to establish a New Ireland colony in the area between the Penobscot and St. Croix Rivers, but the project was nullified during the Treaty of Paris negotiations. For these reasons Morse may have felt his "information from the Southern is much more particular than from the northern states." He warned his readers that although the map of the southern states was the "most accurate yet published respecting that country," the map of the northern states was "compiled principally by the Engraver, from the best Maps that could be procured"; it was chiefly designed to give the reader an idea of the relative situation and comparative extent of the several states and countries comprehended within its limits.[42]

If his political map of New England was only relatively accurate, Morse nevertheless provided discrete natural histories for each state, relying on secondary sources such as the third volume of Jeremy Belknap's *The History of New Hampshire* (1792) and Samuel Williams's *A Natural and Civil History of Vermont* (1794), each of which described aspects of the environment enclosed within putatively stable political borders.[43] By choosing political units for their natural histories, these early American writers followed the conventions of early modern chorography. Chorography was the practice of surveying a place on some local or regional scale (as opposed to the global scale of geography or the universal scale of cosmography). It identified only those features that coincided with and were limited by some given outline, which, both for the sake of convenience and partiality, was often the outline of a property, town, colony, empire, or nation. This contents-in-a-container approach to natural history was so typical that it effectively reinforced and naturalized political geography. The chorographical assumption that environmental and political geographies were equivalent in extent was implicit in the titles of numerous geographical or natural history surveys published in the eighteenth century, such as Charles Brockwell's *The Natural and Political History of Portugal* (1726), Peter Browne's *The Civil and Natural History of Jamaica* (1756), and Thomas Jefferys's *The Natural and Civil History of the French Dominions in North and South America* (1760). These "natural and political histories" suggested that ecological habitats coincided with political borders—an approach that was undoubtedly pragmatic for capturing finer details that would be lost at other scales or for claiming sovereignty over some range of natural resources. But for the purpose of isolating many other salient environmental features, this kind of political ecology was fundamentally arbitrary: first, because climate, weather, soil conditions, seeds, animals, and birds tended to range beyond the fiat borders of states; and, second, because it could imply that political borders determined natural conditions. This partisan approach to natural history was particularly glaring in places where property lines or political borders were very new or disputed, as was the case in Vermont when Williams published his book and over a long period in the broader region encompassing New England and Nova Scotia. Whatever their motivations, on paper writers and cartographers frequently conflated the natural history and political history of the region, either as a whole or in its constituent parts.[44]

Aware of such slippages, in 1787 South Carolina physician and historian David Ramsay suggested to Morse that he try using climatic geography, rather than state and national borders, to organize information on natural history. He asked if it might be "better to describe the natural history of the globe by zones than in the common way of following the political division? e.g. to describe the animals minerals face of the country climate & etc. of the torrid zone & of the other zones would be in my opinion better than to describe the animals minerals & etc. of Egypt Abyssinia & etc."[45] Perhaps in response to this advice, in the second edition of *American Geography*, Morse offered "a natural history of the globe" organized by torrid, temperate, and frigid zones and lines of latitude, including thirty climates of "our habitable world" and a list of cities or regions that fell into each division. However, even in this format, Morse insisted on a tidy climatic distinction between New England and Nova Scotia that simultaneously reinforced their political boundaries: The former was in the seventh climate with Rome, Constantinople, and the Caspian Sea; the latter, joined to the rest of British North America, was in the eighth climate with Paris and Vienna. These groupings comparing the weather in coastal North America and inland Europe were not only facile but also largely wrong or at least arbitrary; in any case, this apparently alternative view had little impact on the book's overall organization. In subsequent editions, Morse retained his narrative entries on the climate and natural history of individual states, colonies, nations, and empires, demarcations that assumed a perfect overlap between natural and political geography.[46]

Such reifying overlaps were also common in the chorography of larger units such as continents, hemispheres, and transcontinental regions. In his other geographical works, Morse leaped from the national to the hemispheric with little comment. He followed *American Geography* with the publication of *The History of America, in Two Books*. According to the first volume, America was a genuinely "vast country": "one of the four quarters of the world," which extended from 80° North latitude to 56° South latitude and contained "almost all the variety of climates which the earth affords." This broad geographical description implicitly formed the background to the much narrower focus of the second volume, "A Concise History of the Late Revolution," on the political and military origins of the United States from 1763 to independence. Drawing on the idea of natural boundaries, French traveler Constantin-François Volney's widely read gazetteer of the early United States took Morse's work as authoritative, but Volney asserted that the natural and political geography of individual states within the nation were already interchangeable. Arguing that these relationships could be further generalized at the regional scale according to climate, Volney divided the United States into four climates corresponding to natural and political regions: cold, middle, hot, and western. The "coldest climate" was coincident with the outer collective boundary of the New England states, a "natural boundary" that could be "traced by the southern side of Rhode-Island and Connecticut on the ocean, and interiorly by the chain of hills, that furnishes the waters of the Delaware and Susquehanna." However, Volney believed the region's northern limit was only temporary because he also posited that "the most simple idea" of the nation as a whole could be encapsulated in terms of the major bodies of water that surrounded the North American continent. "In an age when the advantage of natural boundaries are so well known," he argued, "we can scarcely question, that these will sooner or later form the limits of the country, as they are so distinctly marked." The complementarities between the United States and the watery edges of the continent seemed to him so strong that the nation would inevitably extend the northern boundary of New England.[47]

The naturalist Thomas Pennant manipulated the environmental and political geography of northeastern North America on a still grander scale. In the 1760s, Pennant had published zoologies of Britain and India. In the 1770s "when the empire of Great Britain was entire, and possessed the northern part of the New World," Pennant was working on a "sketch" of North American animals. The American Revolution abruptly curtailed his work. Unwilling to altogether quit the project he repackaged his research material as the natural history of a transnational northern region and published it in 1785 with the title *Arctic Zoology*. This book was organized as an annotated checklist of species that integrated other naturalists' work on the circumpolar Arctic, Europe, and Asia with that of the "northern part of the New World" as far south as the Carolinas. Most editions included two global maps of this idiosyncratic Arctic. The first map, which centered on the Atlantic Ocean, pictured North America from the North Pole to Long Island (and also showed the Bahamas). In this view, New England was usurped into the natural history of a nearly hemispheric zone.[48]

These examples demonstrate that eighteenth-century natural histories and geographical works often pretended that climate coincided with political geography on multiple scales. But it should be noted that, although writers seemed to be describing the environment of an entire domain like New Hampshire or the Arctic, in fact they focused only on selective features of the climate and biota. To represent the natural history of the whole place, they presented limited samples, accounts of remarkable weather events and

FIGURE 1.6 Pennant's *Arctic Zoology* included this idiosyncratic map of the Arctic. From Thomas Pennant, *Arctic Zoology*, 1st ed. (London: Henry Hughes, 1785), Vol. 2, Supplement, 3. Courtesy of Biodiversity Heritage Library.

wild areas, or lists of the most common seasonal conditions and abundant endemic species. As such, these natural histories offered emblems, microcosms, or synecdoches of a given environment—rather than comprehensive inventories of it.

Many naturalists freely acknowledged the necessity of such preferential, representative sampling. In the title to his catalogue of wild and cultivated plants growing in Massachusetts, for example, the botanist Manasseh Cutler identified the state only as "this Part of America." He had initially planned to produce a full list of plants in Massachusetts. In 1780, when he solicited the support of the Harvard Corporation for his botanical study, he proposed "to investigate the botanical character of such Trees and Plants as may fall under my observation, which are indigenous to this part of America, and have not been described by Botanists; also to make out a Catalogue of those which are found growing here, but have been found in other parts of the World, and therefore need no botanical description; and of such as have been propagated here, but are not the spontaneous production of the Country."[49] Rather than an inventory of all native and domesticated flora, Cutler published a catalogue titled "An Account of Some of the Vegetable Productions, Naturally Growing in this Part of America, Botanically Arranged." This 1783 work described "some" plants confined to Cutler's neighborhood on the north shore of Boston, listing those species he had managed to find "growing within the compass of a few miles; except a small number that happened to be noticed at a greater distance." He also noted that the indigenous plants he collected were probably as numerous "as any country produced in a similar climate"—a climate whose character or geographical scope he did not describe. At least he acknowledged the limitations of his published research, whereas Jacob Bigelow, one of Cutler's successors, argued in his *Florula bostoniensis* that the vegetation of the city of Boston should "serve as a tolerable specimen of the botany of the whole New England states."[50]

Conversely, some natural histories or surveys provided overviews of the most widespread climatic phenomena or plant and animal species. Thomas Caulfeild, an early eighteenth-century governor of Nova Scotia, wrote a detailed report of the natural resources in various sections of the province, but his description of the vegetation on French-controlled Île Royale was particularly succinct: According to him it was "intirely a Rock covered over with moss." In New England, Belknap's botanical survey offered "briefly to take notice of such as we are endowed with the most"; Williams decided not to "enumerate" any plants except for those that were "the most common and useful" in Vermont. In a similar rhetorical move, Arthur Young's phrase "north of tobacco" made the presence of tobacco a metonym for the South and its absence a metonym for the North. This method was remarkably close in approach to the renowned Swedish naturalist Carolus Linnaeus's classification of climate zones according to commonly occurring indigenous species. Linnaeus's hybrid model integrated the providential distribution of plants and animals that survived the Flood with classical zones based on latitude. Initially, he proposed five climate zones: the Mediterranean, Australian (Ethiopia to southern Africa), Oriental (Siberia to Syria), Boreal (Lapland to Paris), and Occidental (Canada to Virginia, China and Japan); later, he added a sixth based on altitude, the Alpine zone (all mountainous regions). In this scheme, the close relationships among the biota of northeastern North America, northeastern Asia, and the Alps meant that New England and Nova Scotia's climate was a combination of the Occidental and Alpine zones.[51]

In his agricultural dictionary *The New England Farmer* (1790), Samuel Deane also played with these kinds of rough equivalences, though he encouraged promiscuous investigation of anything in the local environment that might catch an observer's gaze. Deane argued that the New England farmer "must govern all his schemes of management by the peculiarities of the climate," which could best be described by "our countrymen, rather than by strangers." "Farmers on the east and west of the Atlantick" required specific information about the local climate "because that which proves successful in one, will not do so in another." But while stressing the specificity of local climates, he never explained what geographical definition distinguished one from another or strangers from countrymen. Farmers situated west of the Atlantic could mean farmers throughout the entire New World. Deane may have had the new United States in mind when he referred to his countrymen, but the book was ostensibly addressed to farmers in New England. Was New England a region? His technical dictionary provided no direct answer and no discrete entries for the terms "New England" or "region."

Deane's entry for climate was also vague about geographical distinctions, partly because most of it was a direct quotation from the chapter on climate in a Scottish agricultural manual, which itself borrowed from Samuel Johnson's definition, explaining that in the British Isles, "a very small distance sometimes makes a very great difference in climate." Deane noted that, with the exception of geographers who used the term climate to designate the space between lines of latitude, climate was "often used less accurately, to signify a region, or large tract of land." Deane deployed climate in both these senses throughout his text. Because New England's climatic geography was so complex, Deane called on farmers to record the weather and seasonal changes in their own vicinities and, at the same time, wanted to recruit "some attentive naturalist" who could sort such climatic records according to the "42nd, 43rd, 44th, and 45th degrees of latitude." Eventually, he hoped, these empirical data would be compiled and analyzed to conclusively characterize the patterns of what Deane persisted in referring to with the ambiguous collective term "our climate."[52]

Although Deane and other writers increasingly insisted on the necessity of reporting on local conditions, they also studiously avoided defining the precise domain of this localized expertise. Cutler's and Bigelow's representative samples were concentrated in small areas; Caulfeild, Belknap, Linnaeus, Williams, and Young restricted their attention to prevalent species. Both types of selective representation of nature presumed the interchangeability of the selective and the general, the part and the whole. As historian D. Graham Burnett has written of imperial maps, surveys of early American nature sometimes "conjured omniscience" but were more often "merely a handful of views from here to there"—a collection of microcosms or exemplars from which the environmental conditions of broader areas could be inferred or extrapolated.[53]

Nevertheless, their turn to empiricism was a significant departure from the much broader inferences that early modern naturalists had made about New World climate zones based on latitude. As latitude was increasingly perceived to be an undependable, simplistic, or false guide to the local climate, some observers were motivated to discover other factors besides geographical coordinates and the sun's influence that might account for transatlantic temperature differences and extreme seasonal conditions. Settlers and mariners were especially attentive to the powerful winds that originated northwest and south of New England and Nova Scotia. William Wood believed that New England's "more Southerne latitude" helped explain why its "summers be hotter than in England" but it was "only the Northwest winde coming over the land"—not latitude—that caused "extreame cold weather" in winter. This was the same sort of "NW wind" that John Winthrop, Sr., experienced during his first winter in early seventeenth-century Boston. These frigid winds could be so sharp as to cut one's "face like a razor," but when they collided with warm tropical air carried by "the South winde coming from the sea" the collision moderated winter conditions along the coast and intensified summertime heat and humidity throughout the region.[54]

Through the device of a conversation among sailors in a London pub, Joseph Moxon summarized seventeenth-century theories about the geographical causes and effects of winds on air temperatures: "experience tells us, that all Land-briezes are colder than those that come from Sea, and therefore it may be cold about Greenland, because of the Land, and yet warm under the Pole, where the Sea is open." Exploration in the Midwest suggested that continental winds were made colder in winter because they passed over the frozen "vast fresh-Water Lakes," a theory some naturalists later disputed as a "vulgar error, often retailed by geographical writers and adopted by unthinking people." Thinking people proposed two alternative explanations. One was that northwest winds passed over the snow-capped, forested Appalachian ridge that lay west of the region. Another was that the region's coldest and hottest air temperatures originated at high altitudes and were pushed downward by the wind. Two Yale professors touring the coast of Maine in October 1799 believed these to be plausible explanations for every aspect of New England's extreme weather, "particularly the peculiar degree of cold experienced in this country." Naturalists combined these principles with their knowledge of Atlantic trade winds and applied them to understanding climatic anomalies across the region. Throughout the year, the interaction of strong winds from multiple directions produced violent storms and hurricanes, as well as thick fogs that blocked sunshine in coastal areas. On the other hand, some believed that insufficient wind pressure created malignant fogs, the coastal counterpart of malarial air that hovered above stagnant ponds and streams. Benjamin Franklin synthesized a variety of such observations and added further dimension to them in his description of the dynamics of the Gulf Stream, which in turn helped northern naturalists explain why air temperatures were sometimes higher in Nova Scotia than in Connecticut, even though the former was "so much further north."[55]

These empirical attempts to understand the regional climate's distinctive character represented a new emphasis on the geographical variability, diversity, and complexity of American environments. Even so, naturalists, geographers, and other writers never entirely abandoned the ancient model of climatic bands based on latitude. Careful attention to the wide variety of idiosyncratic accounts of the climate in New England and Nova Scotia reflects the fact that, over the course of the early modern period and through the beginning of the nineteenth century, the idea of climate zones as uniform stripes existed in tension with highly specific descriptions of conditions in particular places. As a result, learned elites became increasingly uncertain about the geographical distribution of temperate climates and of northeastern North America's situation in relation to them. The true nature of the region's climate eluded universal, enlightened consensus, remaining a vigorous subject of transatlantic debate.

2

Transatlantic Networks and the Geography of Climate Knowledge

My great pleasure lies in the intimacy & delicacy of friendship and in the pursuit of Science. Happily for me no events of war or politics could ever deprive me of the former or force me from the latter: and I shall never forget the happy hours we formerly spent together in these enjoyments. I heartily rejoice in all your good fortunes and hope the time is come when Great Britain and America will be as happy in alliance and mutual affection as we were in our final hours of friendship and philosophy at Bradford. And if you should have any commands on this side of the Atlantic you will oblige me much to put it into my power to do you any service here.

SAMUEL WILLIAMS TO BENJAMIN THOMPSON, April 20, 1787

SPEAKING OF THE Northeast's natural history in 1790, William Dandridge Peck asked Jeremy Belknap, "How grossly ignorant of this Country is the European World?" Peck supplied his own answer: "the European world, even England is still unacquainted with us." As their friend Manasseh Cutler wrote, the Northeast was like a gray area between Quebec and the Southeast. These places, "beside the attention paid to their productions of some of their own inhabitants, have been visited by eminent botanists from Europe." The area that "lies between them," however, "seems still to remain unexplored." They decided that they had to publish their own accounts of the "Northern States" because, Belknap asserted, "America can best be described by those who have for a long time resided in it."[1]

As the many descriptions of northeastern North America's climate discussed in the previous chapter make clear, in the early modern period and through the end of the eighteenth century, numerous official or private sponsors had commissioned and published travelers' reports about the region.

Despite these efforts, learned residents in New England and Nova Scotia doubted that visitors had done sufficient, worthy, or reliable fieldwork. They disparaged foreigners' published accounts as erroneous or superficial, especially impressionistic descriptions of harsh weather or poor resources in the region. Just as often, they entirely omitted mention of collecting or surveying expeditions pursued over two centuries, relating their own observations of the Northeast's climate and landscapes as if they were wholly unprecedented.

Like most early scientific Americans who wrote about climate or climate change, Belknap, Cutler, and Peck were what would be anachronistically understood as dabblers in applied environmental sciences. Each man was passionately engaged in the sciences of natural history, which encompassed the closely linked studies of climate, vegetation, animals, minerals, and innumerable other elements of the natural world. Belknap was working on a new edition of his *History of New Hampshire*, including an additional volume describing the state's climate, based largely on his friend John Wentworth's journals of his pre-Revolutionary hikes; Cutler had recently published a botanical survey of Massachusetts; and Peck was an amateur entomologist who later was named Harvard College's first professor of natural history. Before the rise of professional science practiced within academic and state-sponsored institutions in the second quarter of the nineteenth century, the credentials to discuss climate and other aspects of the natural world depended as much on social status in and connections to local, imperial, and transatlantic networks as on intellectual abilities, training, or unique discoveries. A naturalist with expertise in climate could be anyone with the leisure and interest to study and write about it. There was no single authority or institution on either side of the Atlantic that coordinated or adjudicated knowledge about local environments. Even the Royal Society, despite its stature and influence, did not dominate the field practices of natural history. Early American naturalists, like their counterparts elsewhere, were officials, clergymen, physicians, lawyers, merchants, or large landowners—governing elites and learned gentlemen and women who, in addition to their primary occupations or other pursuits, also studied, described, and managed local environments.[2]

By the eighteenth century, when scientific approaches to managing the environment became a prime topic of conversation among European and colonial landowners, their engagement in debates about natural history, including the character of local climates, was also a form of cosmopolitan sociability. Learned elites communicated with one another as social, commercial, or political opportunity allowed, forming a loose community constituted mainly of long-distance relationships that were maintained through correspondence and occasional visits. These networks structured scientific patronage, fieldwork, publications, and specimen and instrument exchanges in the early British Empire, and they continued to do so for North Americans through the end of the eighteenth century on all sides of new political loyalties and boundaries. Local elites with competing political and economic interests remained connected through these circuits, which often cut across ideological divisions or proprietary claims to knowledge. Political tensions sometimes irreconcilably divided naturalists. Potentially lucrative technical innovations or economic plants were protected from publicity. Nevertheless, if naturalists in New England, Nova Scotia, and Britain frequently disagreed on any number of scientific or political ideas, especially after the American Revolution, they made sure that their intellectual and affective connections across national and imperial borders survived what Harvard Professor of Medicine Benjamin Waterhouse called the "unnatural" political break with Britain. Waterhouse had supported American political independence but in 1793 he wrote to Royal Society President Sir Joseph Banks: "I really wish to see (as you express it) the claims of consanguinity renewed which subsisted before the war, especially among the men of Science, who if I mistake not, ultimately govern both countries."[3]

As this exchange between Waterhouse and Banks makes clear, the American Revolution did not destroy decentralized networks of scientific exchange. War and independence briefly disrupted or rearranged the geography of communication channels between learned elites in the Northeast and their contacts abroad but these relationships remained largely intact through the revolutionary era.[4] Most scholars have studied the history of early American science and its institutions either between the founding of the Royal Society in 1660 and the beginning of the Revolutionary War or after 1783, neglecting continuities across colonial and national history or the regional and transatlantic ties that further intensified among naturalists in Britain, its empire, and the new United States in the age of revolutions.[5] The practice of natural history in particular hinged on the maintenance of links between dispersed communities that were not exclusive to any one city, colony, region, nation, or empire. Naturalists in London, Edinburgh, Leiden, Madrid, and Paris relied on dispatches of local observations or specimens from American naturalists like Waterhouse, who, in turn, remained crucially dependent on the material and symbolic support of metropolitan patrons.[6]

The resilience of these links was due to the fact that they were largely extrapolitical and extra-institutional. In the British Atlantic world it was mainly voluntary societies and interested individuals that subsidized the development of natural history by sponsoring field surveys, specimen collections, and publications on the diverse local environments of the Americas. Learned elites in the Northeast constituted one regional hub in the increasingly global community of naturalists, travelers, and improving landowners taking shape in this period. But local elites were not disinterested participants in transatlantic science, particularly in debates that concerned northern environments. By highlighting the ignorance of outsiders and valorizing their own direct experience in the region's environment, local elites appointed themselves as experts on its climate, flora, fauna, and soils. By circulating their reports to a wide audience of learned elites across the Atlantic world they would gain recognition as authoritative practitioners of natural history and perhaps, they hoped, be able to better control outsiders' perceptions of the region's climate.[7]

Inquisitive Minds

William Dandridge Peck expressed a common sentiment. The northern colonies' major problem, complained a writer in Nova Scotia in 1791, was the dearth of accurate information about the region's nature. There was "no part of the British dominions," he wrote, "whose soil and climate have been more misrepresented than those of Nova Scotia, which has been injurious to the Province in many respects." In the 1790s, Benjamin Waterhouse had "a strong desire" to communicate "the peculiarities of this new world" to "the European world," writing to Joseph Banks, "that even England . . . was still unacquainted with us." Samuel Williams, Hollis Professor of Mathematics and Natural Philosophy at Harvard College from 1780 to 1788, wrote that, although the "natural productions of this continent, have been one object of general inquiry" few had been attempted by settlers in the Northeast, who were "obliged to depend upon transient and partial accounts." Williams integrated the disparate information available from their reports to write a natural history of Vermont. Nevertheless, he believed that "the subject instead of being fully explored, is yet a treasure but little examined." Belknap was less forgiving in his account of New Hampshire's climate, which had as yet been represented only in the alarmist "dreams of European philosophers, or the interested views of European politicians." Europeans' ignorance, fantasy, and neglect must have continued to prevail because, as late as 1817, David Humphrys offered Banks "a few facts respecting the *sui generis* State of Connecticut." Identifying himself as "a native," Humphrys was concerned that while "all sorts of singular and anomalous articles . . . from a mammoth to a mite, are sometimes sent for your inspection and classification from all quarters of the globe," the "new or curious" aspects from his home quarter had long been excluded.[8]

Such self-aggrandizing claims were attempts to assert expertise and to solicit correspondents and patrons in the genteel world of science. Before the nineteenth century, however, early Americans did not typically identify themselves in writing as naturalists or scientists in large part because it was a preprofessional age. Similarly, as the author of an American "glossary of provincialisms" published in 1816 noted, with one minor exception there was no such thing as an "improver": "though this verb is so common in New England, the corresponding noun, improver, is not in use; but we always say, the occupier or occupant of a house, or land. I have, however, once met with the noun improver in the Laws of Massachusetts." The absence of these terms of self-identification reflected the fact that natural history or agricultural improvement—like any scientific pursuit—were generally not recoganized as formal academic disciplines in the Anglophone Atlantic world. They were practiced more like serious avocations, dignified and productive hobbies that complemented landownership or territorial administration.[9]

Naturalists and improvers were elites or aspirants in British or American political, commercial, or church affairs ranging from the aristocracy and landed gentry of Georgian Britain, plantation masters, and the earliest presidents of the United States to university-educated ministers, schoolmasters, lawyers, and the upwardly mobile merchant families of the Atlantic world. In the northern colonies, members of these groups formed learned societies that functioned like social clubs where natural history was discussed. These clubs were modeled on a combination of the Royal Society, English and Scottish agricultural improvement societies first established under the Black Act, and the Bible-study groups proposed by Cotton Mather in his 1710 *Bonifacius: Essays to Do Good*, which envisioned associations composed of "a dozen Families, more or less, of a Vicinity," who would "Agree to Meet (the Men and their Wives) at each others Houses, once in a Fortnight." With more secular purposes in mind, Cutler and other friends of the almanac writer Nathaniel Ames and his family gathered in the 1760s at Ames's Dedham, Massachusetts, tavern as the Thursday-night or Free Brothers' Club, where natural history was frequently the focus of conversation. A decade later, "a constellation of very estimable and talented persons," including Judge John Lowell, later a president of the Massachusetts Society for Promoting Agriculture, formed the Essex Junto.[10]

These societies were a formal expression of the lifelong connections among family members, neighbors, former college classmates, and business associates. At the same time, implicitly or explicitly, members of learned societies deliberately maintained the distinction between themselves and "ordinary" settlers. Just as Mather addressed his essays to particular do-gooders ("magistrates, ministers, physicians, lawyers, schoolmasters, wealthy gentlemen"), Connecticut minister and Yale Corporation member Jared Eliot was encouraged that Franklin and "other gentlemen in these parts, of worth, capacity, and learning" had bought copies of his writings about scientific agriculture, *Essays Upon Field-Husbandry in New England: As It Is or May Be Ordered.* The Massachusetts Society for Promoting Agriculture's handbook of rules and regulations explained that its rationale for existing was that "many persons, not practical farmers, have associated for the purpose of encouraging useful experiments." Whether the caveat "not practical farmers" denoted a matter of fact or a deliberate exclusion, well-regarded men who took only a minor interest in the sciences were regularly appointed as members of learned societies. For example, one man who was invited to join the Massachusetts Society for Promoting Agriculture was surprised because he had "never made any new georgical [agricultural] experiments and the small degree of knowledge I once had, I have since lost by disuse." He assured the Society that it could "derive but inconsiderable, perhaps no advantage" from its affiliation with him because he had "not sufficiently turned my attention to such matters." Nevertheless he understood that his name in the membership lists would enhance the Society's prestige and, because "the gentlemen who compose the Society are so respectable," he promised he would "not fail of paying my regards to them at their next meeting." The minister Samuel Deane made clear that common settlers were a secondary audience for his specialized agricultural dictionary. His primary readership, from which he solicited subscriptions, was "the rich, the polite, and the ambitious."[11]

The last category was well represented in the northern colonies, where strivers like Benjamin Franklin and Benjamin Thompson (Count Rumford) initiated public careers by gaining the admiration of genteel audiences through scientific experiments and inventions. Engagement in the sciences of natural history established or affirmed the legitimacy of such rising individuals. Although Rumford built his reputation by communicating his work on heat and explosives to the Royal Society, he used gardening and nature-surveying projects to mark his early achievements and to substantiate his extraordinary promotion from Massachusetts apprentice to Imperial Count of the Holy Roman Empire in 1792. As a tutor in Concord, New Hampshire, in the 1770s, he joined in London seed orders with the colony's governor (later lieutenant-governor of Nova Scotia), John Wentworth, who invited Rumford to join him, Williams, and engineer Loammi Baldwin on a scientific survey of the White Mountains in 1773. In the 1790s, Rumford was a principal founder of the Royal Institution; when he served as British consul to the prince-elector of Bavaria, his most notable efforts were the prison and military farms and public gardens modeled on Kew that he created in Munich.[12]

Franklin and Rumford focused much of their exceptional energy on experimentation and theorizing, but both were almost entirely self-taught. No schools in North America or Britain offered thorough training in these subjects until the mid-nineteenth century, so everyone was more or less an autodidact. Many American physicians who practiced in the Northeast were educated in Edinburgh, where they would have heard botanical lectures on *materia medica*; by the late 1780s, Harvard, Yale, Brown, and Dartmouth and learned or agricultural societies in Boston, Providence, and Halifax offered lecture series in natural history by subscription. Both Waterhouse (a physician) and Williams (a mathematician) had been active in natural history at Harvard, but there was no designated position in the subject until 1805. Harvard hired Peck, a minister living in New Hampshire, in part because his reputation was impeccable and he was a far less obtrusive personality than Waterhouse, who had been lecturing on natural history since the 1780s and was better qualified.[13]

Peck's appointment was also justified by the recognition the Massachusetts Society for Promoting Agriculture bestowed on him in 1796 for his illustrations and observations of the slug or cankerworms that infested the region's orchards. But as Peck readily acknowledged, his knowledge of most topics in natural history was extremely limited. In his inaugural address, he cautioned the "Friends of the University" that he faced many challenges in attaining "a more perfect knowledge of the natural productions of our own country" and spoke of the necessity of teaching students to "discover & cultivate" interest in the sciences. For addressing the more immediate goal of producing a regional inventory, however, he felt capable of enumerating only birds and insects and was relieved that Cutler, a Yale-educated Congregational minister who was unaffiliated with Harvard but whom Peck considered as "much more experienced in Botanic Science & rural Oeconomy," was taking on the task of surveying local trees. That both Cutler's and Peck's qualifications as naturalists derived from their experience also reflected the informal, clubby culture of scientific practice in the British or Anglophone Atlantic world in which Americans sought, elevated, or reinforced their social status through scientific pursuits.[14]

Traveling to and describing the environments of distant, exotic, or otherwise obscure destinations was another typical path to gaining the knowledge and cultural capital necessary to be known as a naturalist. After all, accompanying James Cook on his first voyage to the Pacific was how Joseph Banks, who occupied the most prominent position in the Anglophone scientific world in the late eighteenth and early nineteenth centuries, initially earned his reputation. But as long as an excursion was publicized in print credibility could also be gained by exploring landscapes close to home, as English naturalist Gilbert White did in his parish of Selborne. The progression of Manasseh Cutler's scientific reputation exemplifies this latter trajectory in New England. Cutler was not wealthy, but his career as a minister throughout southern New England had been fortunate. An indication of this success was his description of a garden in Newport, Rhode Island, "laid out much in the form of my own, contain[ing] four acres, [with] a grand aisle in the middle," and surrounded by espaliered fruit trees, a hothouse, and "curious flowering shrubs." In Newport, where he served as a chaplain for American troops during the Revolutionary War, he also studied medicine and supported Waterhouse's initiatives to popularize smallpox inoculation. After the war, Cutler ministered in coastal Massachusetts and tried to learn as much as his "leisure would admit" about local vegetation on rambles in his neighborhood and more ambitious trips, like a hike in 1784 through New Hampshire's White Mountains accompanied by Jeremy Belknap and vice president of the Massachusetts Medical Society Joshua Fisher. After one of his walks in Essex County, he wrote to Williams at Harvard that he had discovered "a vegetable, the most singular and remarkable production of nature in its fructification I ever saw and to which I can find nothing similar in any author."[15]

Many of the plants he and his friends found on these hikes had been previously identified by Algonquian and Iroquoian peoples and were not limited or native to New England. But once Cutler presented his research to the American Academy of Arts and Sciences, which published his botanical inventory and made him a member, his scientific reputation quickly expanded beyond the coterie of New England naturalists. In 1784 he was also made a member of the American Philosophical Society, and Birmingham's Lunar Society member and scientific patron Jonathan Stokes requested a correspondence with him. In 1786–1787, Cutler was a key figure in organizing and promoting land sales through the Ohio Company of Associates, which purchased 1.5 million acres in the Ohio Country following the Northwest Ordnance. In anticipation of an expedition to survey new settlements on the Ohio River, he went to Philadelphia in 1787, where John Vaughan, secretary of the American Philosophical Society, gave Cutler a tour of the Pennsylvania State House's public gardens, landscaped with support from Vaughan's father, the London-based West Indies merchant Samuel Vaughan. On the same trip, he met with Benjamin Franklin and studied his copy of Linnaeus's *Systema vegetabilium*; and he was introduced to Benjamin Bartram, Benjamin Rush, and François Alexandre Frédéric, duc de la Rochefoucauld-Liancourt, who was traveling the Eastern Seaboard. Soon Cutler's correspondents included Banks, French botantist Antoine Laurent de Jussieu, and head gardener Jean-André Thouin of Paris's *Jardin des plantes*, as well as prominent naturalists in Germany, Switzerland, Italy, Sweden, and the Danish East Indies. He exchanged seeds and gardening advice with Banks and hosted distinguished foreign naturalists like Constantine Rafinesque, Olaf Swartz, and Count Luigi Castiglioni, an "accomplished botanist" who had been informed that Cutler was "a gentleman better acquainted with Botany, etc. . . . than any other person in the Country."[16]

Colonies of the Republic of Letters

In continental Europe scientific networks of correspondence and specimen exchange tended to be organized by government agencies. Prominent naturalists occupied state-funded posts such as Linnaeus's appointment at Uppsala University or Buffon's directorship of the *Jardin du Roi*. The central location of the botanical garden in Paris, the administrative and cultural capital of France, exemplified the place of natural history not just in that country but more generally in eighteenth-century Europe and European empires. Such institutions monopolized the administration of national and colonial development by sponsoring exploratory journeys and safeguarding rare or valuable finds in botanical collections and gardens.[17] By contrast, British scientific and improving institutions like the Royal Society, the royal botanical gardens at Kew, the Royal Horticultural Society, the Linnaean Society, the Royal Institution, and the first British Board of Agriculture (1793–1822) worked with and for the benefit of national or imperial governments, but their leaders and members were more like expert consultants rather than bureaucrats under the direct control of the state. As historian Richard Drayton has written, Joseph Banks and his patrician cohort styled themselves as "an informal empire of gentlemanly amateurs." They were active in and cultivated affiliations among an array of public and private institutions, the Royal Society, the Board of Agriculture, the Admiralty, the Board of Trade, and joint-stock corporations like the East India Company. The technocratic-philanthropic Royal Institution, for example, grew out of Benjamin Thompson's proposal in 1796 to form "a Public Institution for . . . the application of science to the common purposes of life." In 1799 Thompson, Jeremy Bentham, William Wilberforce, two directors of the Bank of England, sixteen members of Parliament, and other nobles met at Banks's Soho Square apartment to work out the details of how the Royal Institution would function. Like the Board of Agriculture, it was nominally a public institution that promised general benefits to society but was run as a private philanthropy, with membership by subscription and closed meetings.[18]

In eighteenth-century North America, projects or groups for promoting natural history and agricultural improvement were organized on this model, if necessarily with more modest goals and on a smaller scale. Both before and after the American Revolution patronage or control of the practice of natural history in North America was concentrated, but by no means centralized, in Halifax, Boston, Philadelphia, or New York. Instead, natural history societies, surveying expeditions, botanical stations, or publications were usually funded through private or semiprivate initiatives, including those by large landowners located in rural areas outside of colonial cities. Individual improvers' backyard gardens, heated greenhouses, and home libraries housed Americans' most substantial agricultural and horticultural experiments, herbaria, and reference collections. Eighteenth-century northern naturalists were linked to each other and to colleagues elsewhere in the Americas, Britain, and Europe through personal reputation, friendly acquaintances, family ties, or business partnerships. They sought to ally themselves with their counterparts, officials, and patrons in London, Edinburgh, Paris, or Leiden, but such alliances reinforced and were most often forged through regional and transatlantic sociability rather than through formal organizations.

Even when informal associations coalesced into corporate bodies, especially at the end of the eighteenth century, their principal operations relied on already established connections. John Adams assured President of the British Board of Agriculture Sir John Sinclair that local studies of climate, soils, flora, and fauna had "long been desired, by every inquisitive mind." But despite Adams having inserted a clause in the 1779 Massachusetts constitution for a state agency that would provide financing for "a natural history of this country," the initiative had failed. In 1781, Cutler urged fellow members in the newly formed private society, the American Academy of Arts and Sciences in Cambridge, Massachusetts, to sponsor topographical surveys, which were "necessary to furnish materials for a Natural History of the Country, in which we are, at present, very deficient." In the 1780s and 1790s, increasing numbers of learned and agricultural societies formed in North America and, though many of them were incorporated by a charter, all were run as though they were exclusive clubs, even the American Philosophical Society and the American Academy of Arts and Sciences, the only two officially public societies. The Massachusetts Society for Promoting Agriculture, the first agricultural improvement society in the state, was formed in 1792 out of an initiative begun within the semiprivate American Academy of Arts and Sciences; the agricultural society, in turn, sponsored a professorship in natural history and financed a botanical garden at Harvard. In its early decades, both the American Academy of Arts and Sciences and its offshoot organizations could not single-handedly finance or coordinate natural history in the state. Instead, they capitalized on and helped to expand informal networks created by friends and colleagues that carried over from the colonial period.[19]

It was probably through such networks that the Massachusetts Society for Promoting Agriculture obtained copies of *Letters and Papers on Agriculture*, published in Halifax, and *The Nova Scotia Magazine*, extracts of which the Massachusetts society reprinted in its own publications in the 1790s. Some of these extracts from Halifax papers were themselves reprints of columns from the Bath and West Agricultural Society's publications and some were the published advice of writers in Nova Scotia. Natural histories and narrative descriptions or inventories of local resources had circulated in this way throughout the region since the earliest settlements, most of them published in London or reprinted in Boston. Until Jared Eliot's *Essays Upon Field-Husbandry in New England*, the same was true of agricultural advice. Eliot's essays, which appeared in print as a series between 1748 and 1761, were the first published American texts on agricultural improvement. Other items about local natural history circulated publicly in colonial newspapers. In the first issue of *The Nova Scotia Chronicle and Weekly Advertiser* in 1769, the publisher solicited "gentlemen of Experience and Knowledge (of whom there are many in this Colony)," who would be willing to share "their Experiments and Discoveries, as well in Husbandry as in the other Arts and Sciences, by the Channel of his Paper: Such a Correspondence would be a public Benefit." The *Chronicle* mostly reprinted descriptions of climate zones or discoveries in natural history extracted from other newspapers, especially if the texts were authored by a prominent naturalist. The same was true of oral reports delivered at society meetings, for example when the Massachusetts Society for Promoting Agriculture read aloud extracts from Arthur Young's *Annals of Agriculture* and livestock improver Robert Bakewell's "Rules" for breeding "in-and-in."[20]

These patterns of communicating science reflected the fact that the resources, patrons, and membership of learned societies, civic associations, and social clubs that promoted science in various guises were not geographically confined within provincial, regional, or national borders. Instead, they tended to be offshoots of commercial, political, familial, and incipient institutional networks characteristic of transatlantic and imperial relationships in this period. From the beginning of settlement, naturalists in the Northeast actively communicated their agricultural experiments and landscape descriptions to correspondents and institutions on both sides of the Atlantic, nurtured native and exotic plants in private greenhouses and botanical gardens, and encouraged local farmers to develop cash crops and other rural commodities for regional or transatlantic markets. So, for example, when large landowners and gentlemen farmers established the Nova Scotia Society for Promoting Agriculture in the provincial capital and smaller agricultural societies in Annapolis Valley market towns close to the most productive farmlands in the province, colonial authorities and the British Board of Agriculture encouraged but did not control these disparate local initiatives.[21]

These organizations in the Northeast were nodes in the tangled branches of scientific networks rather than the tentacles of some centralized authority that coordinated all scientific activities. When Waterhouse worked to expand Harvard's mineralogical collection in the 1790s, for example, he promised the Harvard Corporation that he would circulate a description of the collection "to every scientific man within my knowledge, in the United States, as well as in <u>Canada</u> and <u>Nova Scotia</u>; and thereby [excite] an attention to this branch of science beyond my sanguine expectations." Improvers in other parts of British North America were also keen to network with their counterparts in the Massachusetts Society for Promoting Agriculture. Hugh Finlay, a merchant and large landowner who settled in Quebec in 1763, was one of the founding members of that province's agricultural society and, after becoming a corresponding member of the Massachusetts Society for Promoting Agriculture in 1793, promised to forward "all such information I may consider worthy of their notice in the course of agricultural experiments in Canada." The exchange would be mutual: The "labours of your Society," he wrote the Massachusetts Society for Promoting Agriculture in 1793, "may greatly promote the Views of a like Society instituted here in 1789.

I trouble you with their publications." In 1818, another member of the Quebec agricultural society wrote to the Massachusetts society requesting permission to reprint extracts from its publications. Since he had moved from central Connecticut to Montreal, where he had by now "lived for many years," he observed that "agricultural knowledge is very much wanted and sought after by the best informed people." By the early nineteenth century, Massachusetts improvers' circles grew wider. Among other connections, one was a corresponding member of the Imperial and Royal Economic-Agrarian Academy of Italy in Florence; another sent wheat seed to the secretary of the Cape of Good Hope Agricultural Society, who hoped to crossbreed New England with Cape wheat varieties.[22]

As a contemporary expressed these relationships in regard to provincial France, learned societies in New England and Nova Scotia aspired to be "the diverse colonies of the Republic of Letters." The small and localized institutions that existed to support agricultural improvement and other natural history practices emerged as a result of such translocal contacts and did not supersede them. Tenacious or newly formed transatlantic connections were reflected in the membership lists of agricultural improvement societies in British North America and the early United States. In 1787, James Bowdoin, a former governor of Massachusetts and founder of the eponymous college in Maine, was elected a member of the Royal Society through the recommendation of Waterhouse and Sir John Temple, the British consul-general in New York. The Nova Scotia Society for Promoting Agriculture admitted as members William Quarrell and Alexander Ochterloney, two military officers from Jamaica who had supervised the deportation of the Trelawny Maroons to Halifax in 1796. In the 1790s, the Board of Agriculture solicited John Adams, John Jay, Thomas Jefferson, George Washington, and other prominent Americans' farming advice and appointed them as members; in turn, agricultural societies throughout North America, including societies in New England and Nova Scotia, nominated improvers in other provinces, states, and countries as corresponding, foreign, or honorary members. Even Adams, a Federalist who championed national development and obstinately resisted most affiliations with foreign scientific institutions as vice president to Washington, accepted his election to the Board of Agriculture on April 11, 1797, the year he was elected president of the United States. And when Adams retired from leading the Massachusetts Society for Promoting Agriculture in 1813, he hoped that the Society would continue to promote the "prosperity of Agriculture and Horticulture in Massachusetts and through the World" rather than the nation.[23]

Washington told Sinclair that his shipments of the latest scientific and agricultural publications from Britain provided not only a "real and important service" to the United States but in turn increased the "fund of political economy, serviceable in all countries." At the end of the War of 1812, Massachusetts Society for Promoting Agriculture President Judge John Lowell believed that the "present perfection" of British and French agriculture was undeniable "even by those who feel the strongest national partialities." He encouraged members of the agricultural society to continue networking with foreign colleagues since "vast benefits have arisen to those countries from the exertions of well informed & opulent individuals associating for the purpose of collecting and spreading the improvements made in different parts of those countries in the important science of agriculture."[24]

In part because of their informality and dependence on individual investment, learned and agricultural societies in early America that were entirely private, such as the Boston Philosophical Society (1683–1688) and the first Connecticut Society of Arts and Sciences (1786–1790), were short-lived. Less than a decade after its founding in 1785, the private Philadelphia Society for Promoting Agriculture was "greatly in decline." Eventually, both of these societies decided to follow the plan of the Massachusetts Society for Promoting Agriculture, which was publicly chartered by the state but restricted "the management of the affairs of the Society to a small, select number of members," Timothy Pickering wrote admiringly of it. The Massachusetts Society for Promoting Agriculture and the Connecticut Society of Arts and Sciences—which was granted a public charter by the state of Connecticut in 1799—outlasted most learned societies formed in the eighteenth century. Still, their limited reach and interest in dispersing rather than centralizing scientific activity and knowledge was attested by the formation of numerous smaller societies in response to the Massachusetts Society's encouragement in a circular sent to town leaders across the state.[25]

Paradoxically, while most early American elites were engaged in some aspect of natural history or agricultural improvement, early proposals for national learned societies never gained sufficiently widespread public support. George Washington's idea for an American Board of Agriculture was a nonstarter. In 1788 Virginians circulated a proposal to establish an Academy of Arts and Sciences of the United States of America based in Richmond, but the project, which would have been underwritten by French capital, never materialized. As Washington wrote to Sinclair in 1794, it would "be some time, I fear, before an agricultural society, with congressional aids, will be established in this country. We must walk, as other countries have done, before we can run. Small societies must prepare the way for greater." However, the failure of national institutions for science in the early United States reflected the new and relatively weak federal government rather than the lack of scientific activity in the nation. In New England, the American Academy of Arts and Sciences, the Massachusetts Society for Promoting Agriculture, and the Kennebec Agricultural Society in the District of Maine made up one cluster of small societies promoting science. Local and transatlantic networks continued to structure scientific activity through the turn of the century. As Cutler explained to a patron in England in 1805: "the present instability of our government forbids the hope of a botanical garden in Washington, and the legislature are far from being disposed to encourage improvements in science. At Cambridge, however, we have the prospect of such an establishment. The plants you mention for Washington will be gratefully received to enrich this garden, and the aid of the friends to such establishments in Europe is earnestly solicited."[26]

Brethren in Science

Although learned and agricultural societies proliferated after 1783 in the early United States, these associations were outgrowths of the informal webs of connections among elites that originated in and were typical of the British Empire rather than institutional novelties inspired by the American Revolution or the institution building stimulated by national independence. The flourishing of the sciences of natural history and the intensification of transatlantic networks of local elites in the third quarter of the eighteenth century reveals the persistence of a colonial sensibility in regard to scientific patronage and the tenacity of pre-Revolutionary relationships. Samuel Williams was made a foreign member of the Palatine Meteorological Society, for example, through the recommendation of his lifelong friend Count Rumford's recommendation to the founder of the society, the prince-elector of Bavaria.[27]

This continuity is strongly evidenced by New England naturalists' attempts to ingratiate themselves with European naturalists, especially Joseph Banks, whose influence over the practice and sponsorship of natural history in Britain and the empire was unsurpassed during his long tenure as president of the Royal Society from 1778 to his death in 1820. Between 1782 and 1783, Samuel Williams sent or received numerous letters from Banks and Rumford, to whom he typically expressed the hope that "with the return of peace," Americans could "again be connected with Britain the country of our friends and ancestors," who would together contribute "to the rise and progress of empire, arts, sciences, and population in America." Not long before the Treaty of Paris was signed, Williams wrote to thank the Royal Society for a recent donation to Harvard, which arrived via Benjamin Franklin: "nothing," he gushed, "could give us greater pleasure than to find that very learned and illustrious body so ready to afford us her assistance at a time when we so greatly need it." He continued,

> The calamities of the war have not destroyed our taste or diminished our zeal in the pursuit of science. They have indeed greatly distressed us as to the ways and means & c. We have not had it in our power to procure such authors and instruments, as we need. But among the blessings of peace we hope to find encouragement and assistance from our friends and a mutual intercourse between the arts and science in Europe and America.

Williams was also heartened that the efforts of the American Academy of Arts and Sciences "to promote the cause of Science are viewed in a favorable light by gentlemen of Science, both in America and Europe." And he assured Rumford that his life's "great pleasure lies in the intimacy & delicacy of friendship and in the pursuit of Science." By 1789, when Williams was forced from his position from Harvard (by some accounts for embezzlement; by his own account because he was a Deist), he wrote to Banks in the hopes that he could lift him out of his exile in Vermont by offering him a post somewhere in Britain or its colonies. "I wish to do all the services in my power for the Society, by collecting materials for useful knowledge in this unobserved part of America," he assured Banks. Williams likely sought a better position anywhere less rural than Rutland as in 1788 he had made the same request to Ezra Stiles at Yale. But in another draft of the letter to Banks, Williams was more emphatic about his allegiance to Britain and the Royal Society: "If it should contribute anything to promote the cause of philosophy my end will be answered: and as I shall ever venerate the British government I wish to announce it to the world by your philosophical transactions." Benjamin Waterhouse's correspondence with Banks was less self-promotional but expressed with as much ardor. Waterhouse had supported American independence but in 1787 he sought the help of the Royal Society in building Harvard College's natural history library. "We look up to the english as our elder brethren in Science & hope to be continually instructed by their labours," he wrote. "Should there be any thing in the line of Natl. History that I could serve Sir Joseph Banks in, he need only communicate it."[28]

Communication and mutual admiration between genteel naturalists and land improvers with otherwise opposed ideological commitments or national loyalties endured through the late eighteenth century inasmuch as they entailed scientific exchanges in which useful, purportedly apolitical knowledge and contacts were generated. In 1792 Belknap delivered an address to the Massachusetts Historical Society in which he praised George III for his patronage of the sciences but whose reign was otherwise "stained with the grossest political errors, and disgraced by the loss of a large portion of this continent from his dominion." Echoing this sentiment, Waterhouse assured Banks in 1793 that "the people of these States regard the Sovereign of England as a monarch preeminently distinguished . . . for his love for the useful and elegant arts." During his second presidential term Washington kept up a correspondence with British agricultural improvers and British Board of Agriculture officers Sinclair, Young, and William Strickland (letters he shared with John Adams and Thomas Jefferson). When they finished their terms, both Washington and Adams looked forward to focusing on farming and scientific experiments. Although Adams sometimes refused to accept membership in or engage in correspondence with foreign agricultural societies, his recusals did not hinder the Massachusetts Society for Promoting Agriculture's efforts to recruit international members and sponsors. In 1801, "having escaped the whirlpool of politics," Adams became increasingly engaged in promoting natural history and agricultural improvement in the American Academy of Arts and Sciences and the Massachusetts Society for Promoting Agriculture, especially as president of the latter from 1805 to 1813, and began to correspond directly with Sinclair and other British improvers.[29]

Waterhouse's involvement with natural history and improvement exemplifies the continuities in scientific networks surrounding the Revolution. Waterhouse's transatlantic scientific connections were personal as well as professional. He first studied medicine in the early 1770s as an apprentice to naval surgeon John Halliburton in Newport, Rhode Island. In 1775, he went to Edinburgh and London, where his main contact was his cousin, Quaker physician and naturalist John Fothergill. Fothergill—Waterhouse's "Kinsman & Preceptor"—introduced him to prominent members of the Royal Society, including George Fordyce, John Hunter, Edward Jenner, and John Coakely Lettsom, who had been Fothergill's tutor. During the war, Waterhouse studied for his medical degree in Leiden, where he boarded with John Adams's sons John Quincy and Charles. After the war, the newly formed Harvard Medical School appointed him as the first professor of the Theory and Practice of Physic in 1782 and awarded him an honorary degree in 1786. From 1784 to 1791, he was also a professor of natural history at the College of Rhode Island (later renamed Brown University).[30]

Despite these professional successes, Waterhouse's career in New England was problematic. Many elite Bostonians who were Federalists, including powerful figures in Harvard's administration, despised Waterhouse because he was an anti-Federalist. In 1812, these differences led the Harvard Corporation to fire him. Thereafter, his working life became entirely dependent on the continued friendship of influential individuals like John Adams, Thomas Jefferson, James Madison, Elbridge Gerry, and Benjamin Rush. Adams was more supportive of Waterhouse in private than in public. When Waterhouse published *The Botanist*, his Harvard lectures on natural history, he dedicated them to Adams. But Adams never bought the book. He was honored by the dedication but he contented himself with his son's copy. As he explained to Waterhouse, the "booksellers in Boston and Salem, who refused to take any of them, disliked the dedicator." The Massachusetts Society for Promoting Agriculture, which underwrote a natural history professorship at Harvard in 1804, did not consider Waterhouse for the position. Adams, who served as president of the society and of the visiting committee that oversaw the professorship, wrote to apologize to Waterhouse that the committee chose "Mr. Peck, who I did not know" instead of "Dr. Waterhouse, who I knew."[31]

Yet if Waterhouse became estranged from northern institutional affiliations, his powerful American friends and transatlantic connections to European botanists in the Royal Society and elsewhere remained crucial to the development of natural history and improvement in New England. Although he had been excluded from the natural history professorship and botanical garden in the early nineteenth century, he was centrally involved in the first plans for a garden at Harvard, which had been encouraged in 1784 by the French monarchy and the *Jardin du Roi* through Hector St. John De Crèvecour, its consul-general in New York. With insufficient additional funding from Harvard or the state legislature, this early initiative failed but Waterhouse continued to raise foreign funding for the project. In 1787, he sent manuscript copies of his natural history lectures to Banks and expressed his interest in renewing the garden project on nine acres in his own backyard in Cambridge, including a hothouse and a chemical laboratory. In 1792, he boasted to Banks that his efforts had excited a growing interest in "the Science of Natural History in general and botany in particular" in "these northern states." Although Waterhouse was never an official Massachusetts Society for Promoting Agriculture member or subscriber to its publications, he kept Banks informed of the Society's activities.[32]

Waterhouse's connections to Fothergill, Lettsom, Collinson, and Jonathan Stokes also proved essential to promoting the sciences of natural history. Before Peck started teaching natural history at Harvard, he depended on Waterhouse for initial contacts and letters of introduction to them as well as to Banks, Sir John Sinclair, William Aiton, Samuel Vaughan, George Fordyce, and Charles Konig, assistant keeper of natural history at the British Museum. Before and after the Revolution, Lettsom sponsored a variety of American individuals and institutions involved in the practice of medicine and natural history, including Dartmouth and Harvard. The latter awarded him an honorary degree in 1790. He was a foreign member of the American Philosophical Society, the American Academy of Arts and Sciences, and the medical societies of Connecticut and New Haven County, and he corresponded frequently with Waterhouse and other American naturalists through the early nineteenth century. Lettsom helped Waterhouse establish Harvard's mineralogical collection and regularly sent him packets of flower and vegetable seeds; Waterhouse sent leftovers to the Massachusetts Society for Promoting Agriculture to be planted in Harvard's botanical garden. "Within five or six years past," Waterhouse wrote to the chair of the society's Visiting Committee:

> I have received nearly 60 papers of Melon-seeds from Dr. Lettsom, with the day of the month, when the melon was eaten; and have distributed them throughout the country to gentlemen & market men; and here send the remaining ten papers to the Trustees. I have received at two different times, within ten years, a large & complete assortment of seeds of all the culinary vegetables of England, from Dr. Lettsom, and have in like manner disseminated them throughout the country.

The Committee depended on such casual gifts because most of the plants that grew in the Harvard botanical garden were the result of donations rather than systematic catalogue orders. Waterhouse also offered the botanical garden "one hundred and two sorts" of garden seeds from Geneva and Marseilles, a specimen of ipecacuanha from the West Indies, and a "peculiarly excellent punkin from the Brazil coast." Because he believed in an "*Ut Spargam* principle"—broadcasting knowledge and materials—and was "pretty constantly receiving seeds from one quarter or another," he told the Committee that he would "be happy in transmitting them through the same channel."[33]

Continuities are also evident in the regional collaborations between local elites on both sides of the new national boundary between the United States and British North America. Elite migrants to Nova Scotia were especially keen to maintain ties to the United States, Britain, and the broader Atlantic world. Many of them were exiles of some sort—Loyalists, Dissenters, British pro-Republicans, pacifists, or opportunists who migrated or accepted land grants in the province during the revolutionary period for a variety of reasons besides political or religious beliefs, from securing the safety of their families in wartime to settling debts or hereditary claims. Once settled in Nova Scotia, they hoped to quickly reestablish connections to their business partners or families abroad. Some achieved a reunion through the most traditional means: the marriage of their children, such as one Cambridge man who in 1790 married the daughter of Jonathan Belcher, a New England planter who served as governor of Nova Scotia and was descended from former governors of Massachusetts and New Hampshire. Waterhouse's first teacher, John Halliburton, fled Rhode Island for Halifax in 1782 and continued to serve in the Royal Navy, but in 1790 wrote to Waterhouse that he wished commercial relations between New England and Nova Scotia could be reestablished. He also wished he could travel to Boston for a visit, but cautiously decided to stay close to his "business" in Halifax.[34]

Improving settlers in Nova Scotia effectively remained part of a broader society outside the confines of the province by affiliating with people, institutions, and publications related to natural history and scientific agriculture. A newcomer from Ireland, dismayed by the political squabbles among Prince Edward Island's great proprietors, "endeavored to unite all parties in the establishment of an Agricultural Society." The minister Titus Smith had been raised and educated in New Haven but in the late 1760s converted to the pacifist Sandemanian faith and evacuated with his parents to a Loyalist camp during the American Revolution. In the mid-1780s they resettled in Nova Scotia with other Loyalists from Connecticut. In the 1790s Smith became active in the Nova Scotia Society for Promoting Agriculture and in 1801 was hired as a provincial land surveyor by the governor. Through the 1840s, he published works on local natural history, promoted agricultural improvement, and maintained a correspondence on these subjects with his family and friends in America, including his sister Rebekah Richardson and his former classmate at Yale, David Humphreys, a former minister to Spain, close advisor to George Washington, member of the Massachusetts Society for Promoting Agriculture, and importer and breeder of merino sheep.[35]

William Almon and his wife Rebecca Byles were part of a more prominent Loyalist family that moved from Rhode Island to Nova Scotia, where they were granted nearly 3,000 acres. In Halifax, Almon served alongside John Halliburton as physician to Prince Edward and John and Frances Wentworth. He was also a founding member of the Nova Scotia Society for Promoting Agriculture. But being elected as a corresponding member of the Massachusetts Society for Promoting Agriculture made Almon "extremely happy." In 1793, he wrote to thank the Boston men who elected him, promising to "promote the views and success of a Society founded upon such a liberal plan . . . I shall certainly avail myself when anything interesting to Agriculture occurs." The following year he sent a book that he had "just received from England" as a gift for the Massachusetts Society's library. Mather Byles, Jr. (Almon's father-in-law and Jeremy Belknap's uncle) worried that family members were constantly "wandr'ng like Gypsies, & playing Puss-in-the-Corner about the Globe" and the extended Almon–Byles clan was "breaking up." Mather had settled in St. John, New Brunswick, half his family remained in New England, and one of his sons moved between London and Grenada as commissary general. In 1801, he wrote to Rebecca in Halifax that "you & I seem to be the only stationary Beings." He joked that they would soon hear that Almon "has made his fortune & gone to live like a Gentleman at Sierra-Leone." But as his son Mather Byles III understood the situation, his travels gave him the "opportunity of enlarging my acquaintance with the world." The relocation and travels of various family members only expanded the geography of their extended contacts in official and learned networks across the Atlantic world.[36]

John Wentworth's biography offers the most compelling example of how relationships among elites in regional as well as transatlantic networks were maintained through discussions about natural history and improvement. Wentworth's status in pre-Revolutionary colonial society was reflected in his class ranking: In 1751, he graduated fifth out of twenty-four classmates at Harvard, where he met Adams, Belknap, and Peck, who remained his lifelong friends. From 1763 to 1767 he lived in London, where his distant relative Charles Watson Wentworth, Lord Rockingham, introduced him to Lord Hillsborough, then president of the Board of Trade. In 1766–1767, he produced a report for Rockingham that justified American protests against the Stamp Act, for which he was granted honorary degrees from Aberdeen, Oxford, and the College of New Jersey (later renamed Princeton). More important, he was appointed royal governor of New Hampshire and surveyor of the king's woods. These roles led him to side with the British in the Revolution, and he essentially reoccupied them in Nova Scotia after 1783, when he was named surveyor general of woods for all of British North America, and in 1792, when he replaced the deceased John Parr as lieutenant governor of the province. Wentworth was made a baronet in 1795. Despite his loyalty, public offices, and promotions in the British Empire, after a brief pause he resumed corresponding with his American friends, frequently exchanging letters with Adams and Belknap and providing foreign visitors with introductions to them. A French traveler Wentworth hosted in Halifax in 1793 was offered "all the good offices in his power while we stayed in his territory; and when we decided to leave it, he also had the kindness to provide us with the best of recommendations for Canada and the United States."[37]

Wentworth's reputation helped his friend Benjamin Thompson—a lowborn schoolteacher from Woburn, Massachusetts—catapult to transatlantic fame as a scientist, philanthropist, and advisor to European monarchs, eventually superseding Wentworth in accomplishments and stature. In 1772 the two met in Concord, New Hampshire, and became fast friends. During the war, Thompson's father-in-law tried to convince him to fight for American independence, criticizing him for maintaining "an obnoxious correspondence" with Wentworth. Thompson protested that, though he and the royal governor had "a long correspondence," it always concerned "purely private and friendly" discussions of natural history, improvement, and other matters "not political." In March 1776, Thompson sailed to England and never returned to North America. By 1783, Thomas Gainsborough had painted his portrait; he was elected a fellow of the Royal Society, and was knighted by George III. By the time Wentworth returned to Nova Scotia, Elector Karl Theodor of Bavaria had awarded Thompson for his scientific and technocratic innovations in Munich and Thompson assumed the name Count Rumford. But in 1796, despite Rumford's continued service to monarchs, the American Academy of Arts and Sciences accepted an endowment of a prize in his name for scientific "discovery or useful improvement" in knowledge about heat or light. It did not bother the Academy that Rumford simultaneously endowed a similar prize at the Royal Society. Mutual regard between members of the American Academy of Arts and Sciences and Rumford continued through the turn of the century. Ten years after Rumford's gift to the organization, the Academy asked Peck to "offer his respect to Count Rumford." Peck deemed himself "unworthy" of the honor of writing to Rumford, but assured the expatriate in 1806 that "in the Northern part of the Union we are beginning to pay some attention to Nat'l History."[38]

Exchanges

Most tangibly, genteel northern Americans contributed to natural history and climate knowledge by exchanging meteorological diaries; botanical, zoological, and agricultural literature; sketches of effective property divisions and hothouse designs; seeds, seedlings, or livestock; and forcing glasses or other equipment for acclimatizing nonnative plants. Like other aspects of natural history, the circulation of these materials was not neatly organized or dominated by formal institutions but instead facilitated by transatlantic relationships, especially after 1783 when British North American naturalists' connections to improvers beyond the region became more extensive. Improvers in Edinburgh, Demerara, and Durban offered Windsor, Nova Scotia, gentleman farmer Charles Ramage Prescott "specimens of wheat from Spain & other countries" in exchange for his "admirable" seed potatoes. Thomas Brewer, an agent for a Boston merchant and Massachusetts Society for Promoting Agriculture member who regularly traded in South American ports, in 1807 promised to return with "Patagonian wheat" for the Society as well as "a great variety of Seeds, roots, shrubs, flowers, &c&c" native to the region surrounding Buenos Aires. Friends, relatives, and colleagues gifted or subsidized each other across short and long distances. Favorite, unusual, or commercially promising seeds, plants, and animals were proof and reward for household gardening or breeding accomplishments, emblems of wealth and privilege, or symbols of affective ties among senders and receivers. When New York naturalist Samuel Mitchell sent Joseph Banks "a small Collection of Plants," he hoped that Banks would accept it "not because they are new, but because they are American and are a token of my respect." Likewise, Peck presented Banks with the transactions of the Massachusetts Society for Promoting Agriculture. Making such contacts was as important as gathering ideas for designing the botanical garden that was part of Peck's duties.[39]

The process of hiring laborers or professional gardeners was another form of exchange that provided an occasion for elites to reinforce their credibility as naturalists with each other. Lord Dalhousie assured Charles Prescott that he would send his groundskeeper Cameron—a man who "knew his business"—to separate grape vines at Prescott's vineyard. When early American colleges or learned societies sought out the practical expertise of professional gardeners to establish their experimental plots they also tended to hire through the advice of previous employers, usually Europeans. In 1781 the newly formed American Academy of Arts and Sciences planted their garden with the help of a recommended gardener from England. Twenty-five years later, the Board of Visitors of the Botanic Garden sent Peck on a European tour where, among other contacts, he was expected to scout for an "Intelligent Gardener" during his visits to Kew, Leiden's *Hortus Botanicus*, and the *Jardin des plantes*. As one committee member wrote to Peck from Boston, "I do not believe one can be procured here at all qualified." Peck ultimately devised his plan for Harvard's garden based on "conversations & consultations" with Gabriel Thouin (son of Jean-André and younger brother of botanist André Thouin), whom Peck endorsed as "a gentleman of eminence in the Profession of ornamental gardening & . . . in the disposition of the imperial & other ornamental grounds in the vicinity of Paris."[40]

In Nova Scotia, colonial authorities not directly engaged in science recognized that the presence of a botanical institution maintained by a skilled gardener was a matter of prestige and diplomatic importance to the empire. Unsure about whether the province's government was a "Votary to Botany," Prince Edward, Duke of Kent, who had lived in Halifax in the late 1790s, asked Banks to write on his behalf to Lieutenant General Henry Bowyer in Nova Scotia. He hoped to "move His Heart that way," that is, to persuade Bowyer to assist with the relocation of the duke's gardener Michael Dalton, who would cultivate the Wentworths' estate and establish a botanical "depot" for acclimatizing plants from Bermuda and the Bahamas. Bowyer obliged the request, assuring Banks that he would "give every countenance & encouragement & shall be happy to be in any way instrumental in promoting the Science of Botany." Anticipating future commands, Bowyer promised continued support for acclimatization experiments in the province, "know[ing] they will tend towards the improvement of Science & for universal benefit." By welcoming the creation of a botanical garden and hothouse in Nova Scotia, provincial administrators hoped to elevate the profile of their fledgling colony. His 1814 obituary in a provincial newspaper stated that Dalton spent the last decade of his life managing the grounds of the British commander-in-chief of North America's summer home. But by the 1820s, the estate was already "relapsing into a state of nature": a "modern ruin" full of "overgrowth" and "vegetable decomposition." Dalton's role in promoting the imperial plant trade and cultivating science in North America lay mostly in the symbolic agreement between his superiors for his transatlantic transfer.[41]

Before the mid-nineteenth century, Harvard's garden and Botanic Institution was the most ambitious effort at showcasing acclimatization experiments in the Northeast. First imagined by Waterhouse in the mid-1780s, the garden began as a small enclosure of "indigenous plants from [local] woods and fields." Waterhouse had also envisioned building hothouses for acclimatizing tropical plants, but Lettsom advised him he would have to "await a future period of wealth and luxury" when the college or the public was willing to invest in it. In 1801, the Massachusetts Society for Promoting Agriculture raised over $30,000 to begin planting seedlings on a six-square-mile tract in the District of Maine, which had been granted by the state legislature; by 1808, nine acres in Cambridge—including Andrew Craigie's four-acre yard behind his summer cottage near the college—were ready for a fall planting. The following year John Lowell sought public support from the Massachusetts House of Representatives for Harvard's costly garden, arguing that a world-class "botanick Institution" in Cambridge was a necessary object of a modern state. Lowell pointed out that "not only every Country but every city of Europe can boast similar establishments equal to the one contemplated here." He hoped that Massachusetts would follow the New York state legislature, which endowed David Hosack's recently established garden in Manhattan. In addition, admission to the garden would be limited to annual subscribers or "occasional" ticket purchasers. These restrictions were supposed to help cover the costs of maintaining and expanding the garden and pay the salaries of the professor and his gardening staff; additional funds would be raised through public sales of seeds, seedlings, plants, soil amendments, and gardening equipment.[42]

The Massachusetts Society for Promoting Agriculture conceived elaborate purposes for the Institution and its educational, ornamental, and commercial gardens. For students of natural history, there would be a demonstration garden planted with indigenous, naturalized, and exotic species; the experimental garden in Maine would be used as a laboratory for plant acclimatization and pest control. In Cambridge, besides research, teaching, and supervising the garden, the responsibilities of the professor included providing guided tours, lecturing on all aspects of natural history to subscribers on request, and enlarging the college's mineralogical collection. Most of all, the Massachusetts Society for Promoting Agriculture wanted the Botanic Institute to inspire public assent to the Society's broader mission, that of spreading interest in the enlightened sciences of natural history and agricultural improvement among northern landowners. The Visiting Committee would ensure that the naturalist hired for the professorship would be "most useful in promoting [the] interests of the University, the arts and the agriculture of the State."[43]

On some counts—and in contrast to the Duke of Kent's abortive plans to integrate Halifax into the growing network of imperial botanical gardens—their efforts were a remarkable success. Peck's tour through Europe, and especially Britain, France, and Scandinavia, had been fruitful: Besides establishing multiple contacts with naturalists and agricultural improvers, he gathered seeds, specimens, and ideas for garden layouts from royal, public, and private estates; hired consulting landscape artists and gardeners; and purchased books for the college natural history library. In the first year of planting, workers harvested and sold hay from the fields intended for the botanical garden. By 1810–1811, they finished a residence for Peck, adjacent to the garden, planted "indigenous forest trees and shrubs," erected a greenhouse and hothouse filled with 350 "exotick plants" (mainly "contributed by friends of the institution who possessed greenhouses in the vicinity"), and established a *glacis* (a sloping earthworks) at the bottom of which was a small marsh for boggy plants. The catalogue of the garden, published in 1818, listed over 1,000 trees, shrubs, and culinary, medicinal, ornamental, and poisonous vines, herbs, fruits, vegetables, and flowers, including imported varieties from the Caribbean, Africa, Europe, and Asia.[44]

But financing all the functions and employees of the Botanic Institution had been difficult from the start. For the garden's outpost in the District of Maine, the Massachusetts Society for Promoting Agriculture "had obtained the grant of land, but not the money" for developing it—a situation John Lowell referred to metaphorically in a report to the Visiting Committee. "The soil of the Garden is ungratefull, cold, & extremely unfavourable for experiments. Its cultivation is unusually laborious." The problem worsened over time. In 1817, Peck was privately complaining to correspondents, "the Botanic Garden here progresses very slowly for want of more ample funds." By the end of the growing season in 1822, the Visiting Committee reported that its budget to pay Peck's salary was exhausted. They eliminated the professorship and hired a "curator" to care for the garden. Nevertheless, by planting the garden, members of the Agricultural Society had partly fulfilled their mission, which they stated at the outset was to provide concrete evidence of the "firmness and perseverance" of local naturalists and improving landowners. While support specifically for the professorship had waned, "opulent merchants" in the region continued to habituate plants from India, China, Africa, and other tropical places to the colder conditions in their private greenhouses and hobby farms, which the Harvard garden continued to supply. In late August 1822, Brown University Professor of *Materia Medica* and Botany Solomon Drowne imagined that Peck must have been "happy" to spend his retirement "residing in a grand garden and surrounded by the vegetable representatives of so many regions of the globe."[45]

Family Networks

Benjamin Vaughan was a prominent figure from a powerful family in British politics, science, and empire. Since the early eighteenth-century his and his wife Sarah's extended families had straddled the Atlantic with investments in estates and plantations throughout Britain, the West Indies, and North America, living for brief periods in each place and maintaining contacts with their local agents everywhere. Benjamin had been a member of the House of Commons, studied medicine in Edinburgh, and associated closely with Joseph Banks, Jeremy Bentham, Joseph Priestley, and Benjamin Franklin. In 1779, he published a miscellany of Franklin's papers in London. A Dissenter, Republican, and supporter of the American and French Revolutions, in 1794 Vaughan fled from London to Paris, where he was briefly imprisoned. After waiting in France and Switzerland for authorization to join Sarah and their five children in Boston, in late summer 1797 they all moved to Hallowell on the Kennebec River in the District of Maine. This northern hinterland of Massachusetts was heavily forested, thinly populated, and far from the people they knew. But as part of an extended family straddling the Atlantic world, the Vaughans were accustomed to bridging distances. While migrants with few resources struggled to settle themselves in the District's rugged coastlands "without friends or neighbors," as one poor man put it, the Vaughans could rely on a vast network of kin, business, and social contacts. The relocation of Benjamin and Sarah's household from London to Hallowell (named after its founder, Benjamin's maternal grandfather) was simply the family's most recent venture in land development.[46]

Benjamin conceded to his brother-in-law in London that going to the northern backwoods meant he would be "resign[ing] all concern with *place* & with active polities." But he believed that this demotion was only temporary. By maintaining contact with an extensive web of connections, he and other resident elites would be the leading agents of improvement in the northern borderlands. Citing the reduction of "Indians and other enemies," the advent of "powerful patrons" like themselves, and the revival of the Kennebec Agricultural Society, first organized by Benjamin's brother Charles in the 1780s, Benjamin predicted that Hallowell would soon be as populous and productive as coastal towns in southern New England. Rather than severing transatlantic connections, the Vaughans' move to Maine (like the Mather–Byles family's move to Nova Scotia) would only further broaden their scope.[47]

The Vaughans kept up an ongoing correspondence with fellow Dissenter William Russell and the natural philosopher Joseph Priestley, both of whom had also emigrated with their families in the late eighteenth century from Birmingham to the United States. Their relationships illustrate how networks of natural history were reconstituted through personal correspondence between friends who moved to new locations. All three families fled England after their homes were destroyed in the riots of 1791. In Birmingham, William Russell had been a successful merchant involved in domestic, European, and Atlantic shipping and drew on his extensive contacts to reestablish himself in the United States. Soon after the family arrived in Boston, Russell's son recalled, "we were waited upon by many gentlemen of the town." Some believed that the Priestleys had ended up in "a bad situation and on bad soil" in Pennsylvania, but William Russell was partially compensated for his lost Birmingham property and settled in Middletown, Connecticut, where he established a successful livestock breeding operation, continued as an Atlantic trader, and managed extensive land holdings in France and Pennsylvania.[48]

If exile in the United States did not impoverish or alienate Russell or Vaughan from scientific or commercial networks, during the years of the Alien and Sedition Acts they effectively "renounced all politics." Unlike Priestley, whose political pamphlets drew criticism from opponents on both sides of the Atlantic, Vaughan did not publicize his views on contemporary controversies and refused to answer frequent queries from his correspondents overseas about his opinions on American, British, or French events. In 1800 William Russell asked him, "With what propriety can any one call this a Land of Liberty? I ever did and always shall think that no Government can be properly esteem'd free where opinion is not permitted to circulate freely & every subject suffer'd to be fairly discussed." Yet both remained active in scientific pursuits, especially agricultural improvement and natural history. According to Thomas Russell, when his father was in Boston, "he was made an honorary member of both the Human and Agricultural Societies while there, the latter of which he frequently attended, and speaks highly of the intelligence and enterprise of its members." The Vaughans, who ran a plant nursery and the Kennebec Agricultural Society, swapped stone fruit scions for the Russells' Leicestershire swine and corresponded frequently with each other as well as with prominent scientific Americans, including George Washington. The families freely discussed apiculture, hemp and flax prices, and Russell's successes in raising English breeds of sheep and pigs in Connecticut. Vaughan's correspondence from Hallowell with "various friends on the continent of Europe" on improvement, natural history, and medical botany was constant and he published several tracts on these subjects, including an edition of the agricultural tract *The Rural Socrates*. He responded to letters sent from Europe discussing the "dangerous tendency of French reformation in Government and Religion," with news about his Hallowell farm. "The returns from this field," Benjamin's uncle in Valance agreed, "are more to be depended on than from the stormy field of modern politics."[49]

Staying in touch by post did not merely preserve their relationships. It was also a means for the Vaughans—improvers in a frontier settlement—to promote the interests of their new location. They reassured their friends at a distance that the family was not succumbing to the northern wilderness and enlisted the support of their social network to recreate elements of an urbane lifestyle in the countryside. On first arrival to Hallowell, Benjamin was "mortified" by the family property "in its present deplorable state," which Sarah compared to "a wretched piece of English common." As a family, they would remake it into the "model of the country." The Vaughans envisioned a "Jamaica fashion" residence with a winding driveway, pipe-irrigated pleasure and kitchen gardens, a greenhouse for indigenous woodland flora, and a "weather-proofed" cellar, financed mostly (and appropriate to the house's design) from profits on sugar grown by slaves in their Caribbean plantations. According to Benjamin's estimate, his family's new living arrangement would cost twice the local average. But clearly their material resources—both financial and technical—ranged far beyond the local. Help came by packet in the form of letters, seed stock, and special equipment, like the forcing glasses sent by Sarah's brother in France. Members of the Royal Society and the Massachusetts Society for Promoting Agriculture forwarded the latest publications explaining chemical and horticultural experiments. The Vaughans promised to return all favors by filling their correspondents' requests for barrels of cranberries, sweet corn, native seed, and news about the progress of their renovations.[50]

Emphasizing how social and family networks structured the study of climate and other aspects of natural history also reveals how eighteenth-century scientific culture involved women and children. Their farm work, gardening, botanizing and other natural history activities otherwise appear largely irrelevant to the institutional origins of modern science. On both sides of the Atlantic, women were tacitly excluded from universities and membership or active roles in scientific societies, despite exceptions such as Stephen Hales' patron and the originator of Kew Gardens, Princess Augusta, and Lady Anne Monson, a botanist who traveled with her husband to India and to South Africa with Kew plant hunters C.P. Thunberg and Francis Masson.[51]

In early America, women were inconspicuous but not absent from scientific culture. Many contributed specimens to museum collections and their works as gardeners, herbalists, healers, or dairymaids were described in newspapers and advice literature. As published writers, they were pseudonymously disguised as male personae. Judith Sargent Murray, who eventually revealed herself as the gentlemanly author of "The Gleaner" column in the *Massachusetts Magazine*, encouraged American women to form intellectual salons and to pursue, like their contemporary Dorothy Schlozer, mathematics, architecture, and mineralogy, "visiting in person, the deepest mines and bestowing minute attention on the several stages of the work."[52]

Family letters are the most revealing documents of women's activities as naturalist or improvers in early North America. That Manasseh Cutler's daughter was an avid gardener, for example, is clear from his praise for her interest in plants. Cutler wrote to her in detail about an impressive garden and private botanical library he visited in Philadelphia kept by Elizabeth Merry, the wife of Britain's ambassador to the United States in the early 1790s, whom he judged to be "quite a botanist." "She has a fine collection of books and a large number of specimens. She appears to understand the science very well, and is a perfect enthusiast in her favorite pursuit. . .wish[ing] to preserve American plants, and to be informed about our vegetable productions." Cutler's daughter and Merry's scientific activities (as well as those of other children, wives, and sisters) become evident if the practice of natural history in eighteenth-century North America is viewed in terms of its wider social context.[53]

From this perspective, letters that might seem merely chatty were instead a form of scientific discussion. Rebekah Richardson's letters to her brother Titus Smith in Halifax fall into this category. Although Smith and their parents immigrated to Nova Scotia during the Revolutionary War, Richardson, her husband, and other siblings remained as small farmers in southern New England. In 1801–1802, when Titus was surveying the interior geography of the province for the colonial government, Rebekah wrote to him about her own discoveries of "natural curiosities" during her "rambles in the woods" near Litchfield, Connecticut. Like Schlozer, Rebekah had been exploring "a number of caves . . . about three quarters of a mile from our house," which, she assured her brother, "would engage your attention for a week." She was "not chymist enough" to know the correct terminology for what she and the boy who came with her to hold the candle saw inside, but managed to fill several pages with careful observations:

> In the highest of these caves I can nearly stand up—the roof is not a regular arch but uneven and the colour is a Bright green. the floor is cover'd with Loose flakes of stone which have fallen from the roof and which are very curious we have picked up and Brought home several roots of Birch and other trees which are turned into stone, some into white flint, others into a substance . . . on the roof of the cave . . . the roots look decayed. there is a continuous dropping of water through the roof which renders the cave uncomfortable. a Little Below this there is another cave which appears much more curious but so wet and Low that I have never attempted to enter it I have however sent in a boy with a candle whilst I Looked in at the mouth the roof is Beautifully arch'd and hangs full of iciles [sic] of stone . . . from the end of which the water drops onto the floor which is smooth as glass and a Little sloping along the Backside of this room there is a deep gutter in which the water stands the depth of some feet.

Like Cutler's hikes in his neighborhood, Richardson used her walks to survey local nature. If the publication of Cutler's touring notes by the American Academy of Arts and Sciences reinforced to the reading public his position as a cosmopolitan gentleman, Richardson's letters to her brother also expressed a learned worldliness, albeit to a much more limited, private audience. Richardson regretted the division of her family between western New England and Nova Scotia. But "the sea that parts us," as she put it, did not prevent her from communicating "epistoleory matter" about American politics, sectarian movements, land speculation, farm news, and natural history to her "philosophical" brother.[54]

The Vaughan correspondence shows their entire family involved in applying natural history to land improvement. Samuel Vaughan Jr. sent Manasseh Cutler a box of various seeds from the East and West Indies, among them "Bengal mountain rice, Bastard breadfruit from the Isle of France," and three kinds of tea to be acclimatized in Maine. In April 1796, when they were staying with Benjamin's brother Charles in suburban Boston, Sarah's brother Henry Bird in Valance organized a shipment of "useful Garden seeds" to Sarah. Bird thanked Charles for his "peace offering" of cranberries and asked Sarah to send a few gallons more when she could. Later that winter, Eliza Bird's note described dinners with American diplomats in London and instructions to "guard the seeds I have lately sent," including hyacinths, daffodils, and "other sweet smelling flowers." The family patriarch, Benjamin's father Samuel Vaughan, sent "Seeds, Tools, Books" from seven different London suppliers and alerted them of an imminent delivery of potatoes and gooseberry cuttings from Lancashire. In 1799, John Vaughan, then secretary of the American Philosophical Society, teased his brother Benjamin that the family should appoint him "travelling Missionary de Propaganda" for promoting their Maine properties as "wilds . . . set off by English Improvements" to everyone in Philadelphia.[55]

Benjamin and Sarah's sons Henry and William Vaughan were tutored in natural history from boyhood. Henry was sent to London in 1801 and wrote to Maine about his impressions of show breeds at the Lewis livestock fair—the cattle were "large but ill shaped"; the sheep, he informed his father, "quite eclipse yours"—and offered to "send some lucern, saintfoin, & nonsuch." In a separate letter, he praised and encouraged William for his improving experiments: "My dear Brother, I have heard with great pleasure of the great improvements you have made for general utility & if I can give you any information I shall with pleasure." Recounting a lecture on the amounts of tannin in the bark of various trees, he recommended that William substitute sumac for hemlock: "as at Kennebec you have plenty of sumach . . . It would at least be worth a trial." Meanwhile, William reinforced already close ties between the Kennebec Agricultural Society and the American Philosophical Society during his own stay in Philadelphia, and again when he returned to Maine and struck up a personal correspondence with Maryland planter and judge John Beale Bordley. The two swapped rutabaga tubers and okra seeds, copies of the *Farmer's Calendar* from England, and recipes for making watermelon syrup. As for the rutabaga, the young Vaughan flattered Bordley that his "father had it before, but the knowledge of your success with it gives new encouragement to its circulation & a name whereby to recommend it."[56]

Sarah Vaughan was not an official member of the learned and improvement societies with which all the men in her extended family were affiliated but she was embedded in the same personal networks of transatlantic society, landownership, and commerce that were sustaining those institutions. These networks rested on the family's cultural and financial capital, which included the Montserrat plantations and slaves she inherited shortly after she settled in Maine. A mutual friend in London gave "credit" to Sarah "for great strength of mind, & much dignity & independence of character," though he expressed concern that "to be so much abstracted from the pleasures of polished & literary society . . . must sometimes prove irksome to you both."

Letters concerning the management of the Hallowell farm and the Kennebec Agricultural Society were sometimes addressed to Sarah rather than to Benjamin. In 1798, a sibling wrote "in haste" to Sarah, urging her to:

> Sow enough of seeds, for seeds to be used the next year—I keep all the sheep, unless you think one Ewe & Lamb at Jones' Eddy will give a double chance for preserving & extending the breed. If you spare now, you will lose one object of serving the country on your own terms. Next year by supplying a 1/2 dozen farmers, the competition will secure your own terms of a further supply.

The repeated use of the phrase "on your own terms" could have addressed Sarah individually or the Vaughans collectively. But it had been Sarah's idea, during their first weeks in Hallowell, to rearrange the property to resemble the master's grounds of a Caribbean plantation. She wrote to Charles with a long proposal for changes:

> If I remain I shall beg leave to divide the inclosure . . . I propose also that the garden should all be in front as you will then be able to watch the laborers & the produce of it from the house & to make an easy fence; having the terrace to divide the garden into an upper & a lower part, or into a vegetable & a fruit part . . . & I propose the road to the house for carriages to go northern & westward . . . & it will be advisable to bring water from the pasture in which the house stands (not the bullock pasture) by means of pipes coming from a reservoir to be made at the spring head, which will at the same time serve as a spring-house.

Throughout this list she interpolated "I wait your direction" or "orders." But Sarah was as determined as her husband to install "a farm . . . which . . . exhibited improved farming & experiments," because, as Benjamin declared in a postscript to this letter, "an English family cannot willingly submit to the privation of antient comforts."[57]

On the Margins

One of Benjamin Vaughan's correspondents wondered whether Vaughan, "after living in the great vortex of cultivated society," would "miss the company of Men of letters" in the rural Northeast. Until the region was more developed, he could at least count on his sons to become "Companions in the walks of scientific philosophy." Engaging in the sciences had a distinctly personal significance for learned elites. It meant being active members of social and intellectual circles beyond the confines of their local community and, in turn, exploiting their membership in long-distance networks to enhance their authority and social status closer to home. Northern elites hoped that their own flourishing farms and manicured domestic gardens would be objects of emulation among locals, who might consequently show them greater deference. Northern elites flaunted their privileged connections to patrons or influential people abroad by importing expensive equipment and unusual flower, fruit, and vegetable varieties, arranging their gardens according to the latest fashions and building hothouses or greenhouses to sustain warmer-climate plants.[58]

A detailed portrait of the Vaughans' neighbor, the midwife and farmer Martha Ballard, provides a basis for comparison between the two households. Martha, like Benjamin and Sarah, grew and traded a great variety of native and imported garden stock (including corn, cranberry, squash, musk melon, chamomile, and strawberry sets) with residents along the Kennebec River on her travels for medical and social visits. But the Vaughans were generally aloof to commoners like the Ballards, and relations between them were strictly formal. Benjamin and Martha may have intersected in their respective work as physician and midwife, and Mr. Ballard once bought flaxseed (which Martha planted) from the Kennebec Agricultural Society. Historian Laurel Thatcher Ulrich observes that "the gardens people cultivated revealed their aspirations," and it is true that a subtle measure of the social distance between these families lay in their gardens. If for Martha, "cabbage stumps, not hyacinths, were the first harbingers of spring," for Sarah, it was the reverse. Benjamin, who was keenly conscious of the symbolism of decorative landscapes, implored his brother in Boston to "buy and beg" for him "as many slips of white currant trees as possible." If the order had to be filled out with red currant bushes, Benjamin asked that not more than in "proportion of one red to three white." Red berries must have appeared too common for his taste; the Ballards enjoyed them from their own garden.[59]

Demonstrating connections to other places was especially important for elites in Nova Scotia, where the British settler population remained low until after they expelled the majority of Acadian settlers in 1755. The so-called New England Planters who resettled Acadian farms and the British naval and military officers stationed at Halifax enlarged regional networks and established the first permanent British occupations of the peninsula. In the late 1760s, a newspaper publisher in Halifax reminded his readers that they enjoyed "the same privileges and immunities which are enjoyed by Yorkshiremen," as well as their closest "fellow subjects in America," whom he called "Boston and New-England men." Regional ties were further strengthened as Loyalists resettled in British North America. The largest number of American Loyalists—roughly 40,000—migrated from the thirteen colonies to the coastal towns of Nova Scotia between 1774 and 1784. Ironically, once in Nova Scotia, the Loyalists tried to distance themselves from the Planters, who were middling farmers with limited ambitions. In 1785 a retired British army officer with a land grant of 500 acres near Annapolis Royal, cautioned a Loyalist from Plymouth, Massachusetts, that the Planters were "bad Public or private characters," with whom he would "ever be unconnected."[60]

Sometimes remaining on the margins of broader social networks was a deliberate choice. Before the mid-eighteenth century, the Mi'kmaq and Acadians operated within kin and village networks and intermarriage between Native people and French colonists was common. A British agent in 1720 observed that they were "firm allies" and mutually "dependant by the tyes of long acquaintance, consanguinity, and Religion," largely insulating themselves from both French and British metropolitan control. These bonds proved resilient even through the expulsions, when British and American soldiers and migrants took possession of Acadian farms. Some Acadians formed new settlements throughout British North America and northern New England and maintained a separate existence. Acadians who survived the brutal expulsions retreated to the edges of British control. Others, who were rejected from the southern New England farms where they had been deported, returned to the Bay of Fundy coast as skilled laborers or tenants working under New England or British landlords.[61] A surveyor assessing tenant farms on the Bay of Fundy coast reported in 1795 that the Acadians kept at "a distance from the Intercourse of others." In another report from the area six years later, the surveyor noted that Acadians wore homespun wool and subsisted on potatoes and freshwater eels. But even they were not entirely self-sufficient or isolated from colonial networks. Each family consumed four pounds of flour per year, purchased with cash from small sales of meat and butter.[62]

Such a circumscribed existence was abhorrent to northern elites. Although they grounded their expertise in local nature, they did so in order to place the region in the mainstream—rather than the eccentric backwaters—of transatlantic intellectual and economic currents. By reporting on the climate and other aspects of the physical world in the Northeast, they forged ties to a geographically dispersed array of government officials, landowners, merchants, speculators, intellectuals, and families interested in all the various aspects of natural history. Sometimes through correspondence, sometimes through collaborations in official or private projects—from land sales and resettlement schemes to the formation of social clubs, learned societies, botanical gardens, and nurseries—learned elites comprised a far-flung community of self-styled expertise. By reaching out to learned elites outside the region, they could "unite Countries together in one great Chain or link," as William Vaughan expressed the common enlightenment sentiment. But while communication links may have unified elites across long distances, they did not generate unanimity on highly contested subjects. On the contrary, the decentralized structure of transatlantic networks allowed for a diversity of opinions to proliferate, facilitating rather than definitively resolving debate about the nature of the northern climate.[63]

PART II

Climate and Colonialism

3

An American Siberia

For this cold climate . . . one ought to be as frozen as the season to exist with any degree of comfort.

FRANCES WENTWORTH, HALIFAX, October 5, 1784

CONSIDERATIONS ABOUT THE northern climate were frequently at the center of immigration and development policies in the Northeast. In Nova Scotia, this was the case from the beginning of its history as a British colony. In 1708, Scottish projector Samuel Vetch marshaled his knowledge of the Northeast's geography into an ambitious plan he called "Canada Survey'd." Capitalizing on the recent Act of Union and his connections to powerful colonial officials in the region, Vetch proposed the British conquest of New France and Acadia. In its aftermath, he recommended the crown pay for the transportation and subsistence of Protestant settlers from northern Europe and Britain, especially Scots. Because his countrymen possessed a "Northerne constitution," he believed they were uniquely suited to withstanding the special physical requirements for life in other northern regions. Indeed, never denying the severity of the local climate, he insisted that the bodies and minds of Scots were specially fit to thrive in it. "Canada Survey'd"—an allusion perhaps to political arithmetician Sir William Petty's influential Down Survey of mid-seventeenth-century Ireland—combined the language of political arithmetic with that of geographical determinism. The region was "exactly calculated for the constitution and genius of the most Northern of the North Brittain." Britons would quickly adapt to northern North America's cold seasons and inevitably exceed the Iroquoian, Algonquian, and French populations in numbers, health, and general prosperity.[1]

The climatic justification embedded in Vetch's proposal was not merely a subterfuge. In part it was born of his terrible experience in Central America's tropical climate during the late 1690s in Scotland's short-lived Darien colony, where Vetch thought he had learned about the nature of empire. Successful plantation colonies in tropical or subtropical climates might seem to offer profitable short-term investments based on bulk agricultural export, but their disease ecology posed formidable barriers to European newcomers' survival. He decided that colonies of settlement should be planted in North America's temperate environments where European migrants would become readily naturalized. It further followed from this logic that northern Europeans who were already desensitized to cold winters should settle in the northernmost colonies. Referring to Scots, in "Canada Survey'd" Vetch pointed out "how infinitely more agreeable" Acadia and Canada's "climate would be to our Northerne constitution than Darien." As Massachusetts and New Hampshire governor (and former governor of the Dominion of New England) Joseph Dudley put it to the Board of Trade in support of Vetch's plan, a "Scotch colony there of five thousand men would find their own Scotch climate and health and a country far surpassing all Scotland."[2]

Vetch and Dudley were only the first in a long line of local elites to promote colonization and devise schemes for changing the demography and ecology of the Northeast that drew explicitly on a climate theory later organized more formally—though not necessarily practiced more successfully—as the science of acclimatization. Both in its early and modern guises this technocratic approach to colonial transplantation and economic development was appealingly simple and seemingly unassailable: Long-distance migration within presumably homogenous climate zones was safe because newcomers arrived already inured to local conditions, as if they had moved only a short distance. Early acclimatizers in New England and Nova Scotia admitted that the winters were long and severe, but argued that the region would be perfectly comfortable to humans, animals, and plants already habituated to the cooler conditions of the northern temperate zone. Because the Northeast was nearly identical to "temperate colder climates" in northern latitudes on the other side of the Atlantic Ocean, they claimed that migrants from Britain, Scandinavia, Switzerland, and the German states would be undaunted by local conditions and skillfully cultivate cold-hardened native and imported seeds, plants, and livestock. Northern transplants would be more likely to make the northern colonies their permanent home.[3]

In a broad sense, since the beginning of colonization Europeans had effectively practiced acclimatization each time new migrants settled in America. In the eighteenth century, however, officials in the northern colonies found the concept of acclimatization an especially useful technique for promoting settlement schemes while at the same time sidestepping transatlantic controversies about population and migration policies. In a period when demographic growth was widely understood to be a favorable indicator of state power and economic stability, settler colonialism in the Northeast came under increasing scrutiny, especially by metropolitan critics who wanted to stem British emigration to North America. A commonplace of early modern state building was that more populous countries—that is, countries with more laborers—generated greater national wealth and security. But open questions remained as to the causes of population and depopulation, the extent to which environmental factors affected population size, and whether colonies contributed to or diminished human resources in the mother country.

Charles Davenant—a bureaucrat and political arithmetician who served the courts of Charles II, James II, and Anne—argued that communal well-being owed more to the quantity of inhabitants than to the quality of local land or air. "Tis' perhaps better that a People should want Country, than that a Country should want People," he advised. The populous Dutch Republic, despite its "bad Harbors, and the worst Climate upon Earth," was nevertheless one of the most prosperous nations in Europe. "Rich Soils not well peopled" were invariably poor and vulnerable to invasion. Yet Davenant acknowledged that peopling rigorous climates was in and of itself a problem of statecraft, particularly for an expanding, far-flung empire such as England's was becoming by the late seventeenth century. How, for example, to entice "Persons of any tolerable Reputation or Capacity . . . to go and reside in Parts so Remote, where the Climate is so contagious"? Although here he was referring to tropical diseases, he also considered the ease of peopling temperate empires. The Ottomans, though "dispeopled" by war, famine, and epidemics in the late seventeenth century, would have no trouble regaining population because "their climate is so good and healthy that they will soon be recruited with People."[4]

Conversely, the difficulty of populating settler colonies in cold climates presented serious political and economic drawbacks, including the possibility that they would be more reliant on external defense and subsidies with little to offer in return. These disadvantages, mercantilists and critics of imperial expansion insisted, argued against pursuing settler colonialism in northeastern North America. If the region's winters were as cold as was widely reported, it would never be a magnet for settlers. If it was instead just like Britain, the northern colonies would be inevitable competitors. Although there were no natural constraints on reproduction in New England's cold climate, William Petty proposed that the region's prolific, able-bodied, long-lived farmers would have far more benefit to the empire if "Transplanted into Old England or Ireland." Aside from protecting Britain's access to the fisheries, it was a fool's policy to encourage emigration to or investment in the region.[5]

Skeptics of northern colonization found it especially useful to compare Georgia and Florida with Nova Scotia, the last mainland colonies to be established in the eighteenth century and financed with state support in what were perceived to be extreme climates. Immigrants to these colonies could expect "to be frozen to death in the cold of the northern settlements; or burnt up in the heats of the southern." In a group of bad acquisitions, which "increased the tracts of northern snow and southern heat," they ranked the northern colonies lowest. Neither subtropical humidity nor cold winters were especially healthy for British settlers, but land in cold-climate colonies had the added disadvantage of being unprofitable. A tract that specifically compared Georgia and Nova Scotia asserted that the latter's climate was so "unfavorable little can be expected of it." Surveying British North America as a whole, it argued for the commonsense view that the Carolinas and Georgia, "though the latest settled and therefore the farthest from the best state of cultivation, yield more valuable articles of trade than the northern colonies" because a variety of crops could be grown and harvested without interruption throughout the year in the South. Meanwhile in Nova Scotia, "the rigour of the climate" in winter "puts an end to all vegetation." So prevalent were these threatening comparisons that they became legend: A Loyalist leaving Savannah for Halifax feared being "frozen to death" in an arctic exile approaching "Nova Zembla or Greenland."[6]

The most vocal and persistent critic of settler colonialism in extreme climates was the inimitable journalist and agricultural writer Arthur Young, who had considered settling in North America in the 1770s. In his autobiography he explained that his ailing, aged mother prevailed in dissuading him, but his enthusiasm for life in the colonies was probably never strong. While he was entertaining the idea of a transatlantic move, he continually published pessimistic assessments of the prospects for empire in North America. In a 1773 tract, Young railed against migration to any colony in continental North America except some patches of land situated between New York and North Carolina, the only region he considered to be genuinely temperate. Even in these colonies, however, the lands at the coast were "a bad country, and in a bad climate" and their western frontiers ran too close to Native territory. "What therefore were inducements for any of our people to leave Britain (except for the class of the unindustrious) in order to settle in the colonies?" he asked. Predictably, he was opposed most of all to subsidies for northern colonies in territories where "the climate is so rigidly severe in winter, the frosts, snows, and fogs so infinitely troublesome, that no people would ever leave Britain to settle, unless government fixed them by encouragements."

In the first issue of his journal, *Annals of Agriculture* (1784), he insisted that settling "the deserts, marshes, and snows of Canada and Nova Scotia" was counterproductive to domestic interests. The northern colonies would always be a net loss to the British Empire, either as a competitor in exports and shipping or as a burdensome colonial dependency. Again and again, he urged curtailing colonial expansion in northeastern North America.[7]

At the same time, over the course of the eighteenth century, elites in the Northeast became more firmly committed to settlement in the belief that increasing the colonial population was necessary to maintaining British sovereignty and a viable agrarian economy in the region. For colonists in New England, wary and covetous of their northern rivals in the fur trade and fisheries, replacing French Catholic populations with loyal, industrious Protestant settlers was an appealing prospect at all times but became a particularly urgent concern in times of war. Invoking the principle of northern acclimatization, proposals such as "Canada Survey'd" addressed these colonial anxieties while avoiding overt confrontation with imperial population politics. By emphasizing the climatic similarity between northern Europe and northern America, rather than migrants' national origins, ethnicity, or confessional identities, colonial officials suggested that, for "Europeans habituated to a northern climate and its modes of agriculture & labour," becoming a settler in the Northeast was really just a form of local migration, a westward change of address within the same domain. In addition, by identifying the region's place in the natural economy of a global northern zone, they might attenuate the long-standing criticism that encouraging northern settlement was counterproductive to the political economy of empire.[8]

Although this strategy of naturalizing settler colonialism in the North worked reasonably well for mobilizing ecological exchanges of hardy domesticated plants and animals, using climate theory as the premise for northern settlement schemes often failed to produce positive results. Nevertheless, the idea of a northern temperate zone of transatlantic migration remained credible throughout the eighteenth and early nineteenth centuries as a form of policy rhetoric, an irresistible first and last resort for local elites determined to settle and develop land.

Population and Depopulation

From their earliest migrations to the region, Europeans interpreted their eventual success in establishing settlement colonies as a kind of cultural versatility, a testament to their own special ability to acclimate to a new environment. Relying on latitude, Martin Frobisher initially predicted that Newfoundland would suit English people because it would be in "the same styll" and "more to the temper of England and the known regions of Europe" than the equatorial regions that Spain had colonized in the Americas. This now antiquated sense of the word "temper" as a compound term for climatic geography, human physiology, and temperament was embedded in a range of reports about the Northeast. Sir Ferdinando Gorges declared that New England was most suitable to the temperate nature of the English. Thomas Morton argued this point by analogy with lions: There were no lions in the region because it was "contrary to the Nature of the beast to frequent places accustomed to snow." Temperate English men and women were obviously a different, more adaptable kind of animal. The same ruggedness was true of the livestock and seeds they imported. As John Smith admitted, "some tender plants may miscarry because the summer is not so hot and the winter is more colde in those parts . . . then we find in the same height in Europe or Asia," but he had also brought hardy seeds to New England with which he "made a garden." Not as frigid as Newfoundland or as steamy as the West Indies, the region's cool temperate climate promised to be good habitat for colonists and their biotic companions.[9]

Before they established permanent settlements in the region, early seventeenth-century explorers also frequently described indigenous people as "of a perfect constitution of body, active, strong, healthfull." Although New England's winters were definitely colder than England's, wrote Bartholemew Gosnold, both his fellow travelers as well as the locals were healthy, which in turn indicated to him that the region had "as healthfull a climate as any can be." John Brereton observed in 1602:

> the agreeing of this Climat with us (I speake of my selfe, and so I may justly do for the rest of our companie) that we found our health and strength all the while we remained there, so to renew and increase, as notwithstanding our diet and lodging was none of the best, yet not one of our company (god be thanked) felt the least grudging or inclination to any disease or sickenesse, but were much fatter and in better health than when we went out of England.

Echoing this correlation between climate and bodily constitution, among the justifications that the Plymouth Company migrants agreed on for settling in Cape Cod was that "the place was likely to be healthfull." These impressions were recorded before the major epidemics that afflicted native groups in the

Northeast beginning with the region's first permanent colonial settlements in 1616 to 1619.[10]

Following other European empires, the English and French practiced a large-scale, ongoing transatlantic experiment in ecological imperialism, introducing and naturalizing Old and New World creatures and plants to nonnative environments. This often deliberate but nevertheless uncontrolled experiment radically altered American landscapes and disease environments, with some benefits to indigenous people but also unfathomably tragic, largely unmitigated consequences. Sustained contact with Europeans exposed Native Americans to invisible imported microbes that caused visible demographic devastation. Indian suffering from contagious disease epidemics was exacerbated by conquest, enslavement, and territorial wars that caused cumulative population loss—in some areas mortality rates were as high as 90 percent including in New England and Acadia/Nova Scotia. Native polities in the Northeast decimated as a result of high mortality from sickness and frontier war were further displaced or reconfigured by the forcible or legal expansion of colonial property, which in itself proved fundamentally incompatible with customary Indian uses of land and coastal resources. In the late seventeenth century, colonist William Hubbard tried to explain the dwindling native populations in the region as a natural fact, maintaining that there had always been few indigenous people in New England compared with warmer regions. Northern climates "were never observed, by any of the first discoverers, to be alike populous with the southern, the land there being less fruitfull, and the winters more tedious and severe." But as modern historians have amply shown, during the first decades of settlement colonizers were keenly alert to the escalation of Native American demographic crisis and geographical displacement as direct and indirect results of European trade, settlement, and agricultural expansion.[11]

By the 1630s, colonists in southern New England could see the marked difference between fields still maintained by native farmers and those places that had been overtaken by "troublesome" underbrush, where a great many Massachusett, Narragansett, and Wampanoag Indians "had died of the Plague some fourteen years ago." By 1634, Winthrop wrote to a correspondent in London that, while the roughly 4,000 colonists in Boston had adjusted to "sharp and longe" winters, "the natives, they are near all dead of the smallpox, so the Lord hath cleared our title to what we possess." Philip Vincent, the author of a celebratory history of the Pequot War in Connecticut, in which nearly 800 Pequot and 100 English died, described the process of colonial seasoning and land seizure on a larger scale. English colonists who had "overcome cold and hunger" in their transatlantic journey were now "assured of their peace by killing the Barbarians" and could live "securely in their Plantations sixty miles along the coast and within the Land."[12]

Newcomers explained both the degeneration of native communities and their own resilience in weathering changed circumstances as results of the natural and divine order. This understanding of English acclimatization was reinforced by their insistence that, despite witnessing Indian susceptibility to disease, the region's long winters suppressed contagions. The merchants of Halifax's harbor boasted that human fevers were kept at bay by "the peculiar Salubrity of our climate." Some thought the bracing winds originating in heavily forested tracts northwest of the region were responsible for the lower incidence of contagious diseases among northern compared with southern colonists. Winds promoted healthy air circulation, as opposed to stagnant, miasmic atmospheres. "Dry, cold and bracing" winds purified New Hampshire's air, making it "salubrious." The local climate was like a tonic for colonists. "Experience" showed Salem, Massachusetts, minister Francis Higginson that there was "hardly a more healthful place to be found in the world that agreeth better with our English bodies." "Many that hath been weak and sickly in old England by coming hither have been thoroughly healed and grown healthful and strong." It seemed not to matter that this was the observation of a man who died a year later from fever.[13]

This quick adaptation to the northern climate augured well for colonial population growth. Because New England's climate was "most apt" for English bodies, it was also "most fitt for the generation and habitation of our English nation," wrote Morton in 1637. In the same year, Vincent estimated that the more than 30,000 settlers in the region by this time "shall every day bee, doubtlesse, encreased." His certainty was justified by the common knowledge that the "English and Scotch" were "good breeders," even more prolific than the Dutch: "a facultie that God hath given the Brittish Ilanders to beget and bring forth more children, than any other nation of the world." Living in "the Air of new England," he argued, only further enhanced this natural propensity and gave the colonists "no small hope of their future puissance and multitude of subjects." In 1672 Samuel Denys expressed a similar habit of thought about European adaptability to the region when he declared that since Acadia's climate was "similar" to that of the mother country it would be "easier to populate than other French colonies."[14]

In southern New England, these observations seemed to colonists to be borne out by natural demographic increase. Over the course of the colonial period, the settler population of the earliest English colonies steadily increased because of high rates of reproduction and low rates of morbidity and mortality. By 1670, southern New England's settler population had exceeded 60,000; a century later, the population of New England as a whole approached one half million, most of it concentrated in Massachusetts, Rhode Island, and Connecticut. By comparison, failed colonization or lower population density and land values were the norm in the seventeenth and early eighteenth centuries throughout much of northern New England and the Acadian peninsula, especially in uplands farthest removed from access to rivers or the coast. Before the nineteenth century, the British settler population in Nova Scotia, the District of Maine, northern New Hampshire, and Vermont increased mainly from transatlantic and intercolonial immigration, especially from southern New England. In the 1760s, after five decades as a British colony, Nova Scotia's settler population had reached just over 15,000. Three-quarters of this growth was due to the relocation of the New England Planters, who occupied abandoned farmlands after the expulsion of 14,000 Acadian settlers. After 1783, migrants from southern New England accounted for the fourfold expansion of land settlement in northern New England. Internal migration within the region partly resulted from the prolific reproduction and social cohesion of southern New England families. Where estates were distributed more equally through partible inheritance, ever smaller subdivisions and the specter of soil exhaustion encouraged settlers to claim lower-priced land in the large, unimproved grants in New Hampshire and the District of Maine; where custom dictated giving first-born sons a double share of land, younger children were more likely to establish households farther and farther from home in the northwestern and easternmost parts of the region.[15]

Another solution to these economic, demographic, and ecological pressures in older colonies was to desert the Northeast. The prevalence of transatlantic remigration from the region's early religious communities has generally received much less attention than the stories of naturalized immigrants. But early Americans knew that permanent settlement was a relatively rare decision. Thomas Hutchinson calculated that between the 1630s—the first decade of the Massachusetts Bay Colony's existence—and the mid-eighteenth century when he was researching its history, "more have gone from hence to England than have come from thence hither." After the Great Migration to New England between 1620 and 1640, as many as one in four erstwhile colonists returned home to England. Of the returnees, one in three were ministers, which suggests that a substantial number of migrants to Puritan colonies considered their New England experience a voluntary exile rather than a definitive move. Between the English Civil Wars and the Interregnum, when English migration to mainland North America declined overall, the smallest proportion of migrants landed in northern ports. Within that period from 1650 to 1656 more New England colonists returned to England than settled permanently in it.[16]

During this period and later on, besides the problems of land hunger in southern New England, the region's climate and stony soils were the secular reasons emigrants most often cited for departing—perhaps because it was easier to blame the weather or landscape rather than the complex social dynamics of the colonies. In the eighteenth century, although some southern New Englanders left to settle northern lands, more moved into the territories of the Iroquois confederacy in western New York and the Ohio Valley, a colonial outmigration that began in the 1740s and intensified after 1783. In 1786, Manasseh Cutler asked Jeremy Belknap for his advice about such a move.

> I have got a new maggot in my head, which sometimes bites pretty smartly. What think you my friend of the Ohio Country? Is it not much preferable to these frozen regions? . . . It appears to me a matter worthy the serious attention of those who have large families to provide for, and have little or no real estate to leave them, that they should provide for them, as they may now have their choice in the best part of the country.

Belknap had to agree. Another mutual friend would have urged colonizing the plentiful lands in the "Eastern Country" of Maine, but if it were his own decision to make, Belknap would "face the same way as the Indians looked for their Heaven, viz., S.W." Once Americans had removed the Iroquois—the only major "inconvenience"—there would be "every inducement" to go to the Ohio Country, including its milder climate ("no wintering of Cattle!" Belknap cheerfully pointed out).[17]

Thus the demographic growth of families in southern New England was reassuring, but it never entirely erased the long-standing fear that cold northern climates had a dampening effect on European population growth and immigration—crucial conditions for the success and expansion of colonial settlements. Failed settlement projects and emigration away from the region in the seventeenth century generated adverse publicity and in turn provoked some of the earliest promotional pamphlets aimed at reviving immigration into the northern colonies. Broadsides about the northern climate became more prevalent as settler recruitment turned into an increasingly competitive business in the eighteenth century. Skeptics of northern colonization—including those who accepted that even the northernmost parts of the region were geographically located "within the Temperate Zone"—reported that the climate had been "found rather unfavourable to European constitutions." Nova Scotia, the author of *American Husbandry* warned, was "dreadful to newcomers." Land companies with interests in other regions extolled "the wide and pleasant fields of North America" in general, but strongly discouraged immigration to the Northeast in particular. Dueling images of the local environment proliferated as wars in continental Europe and the colonies intensified transatlantic and regional migration among refugees and opportunists, particularly during the War of Spanish Succession from 1701 to 1714 and the swelling and shrinking British and French Empires between 1763 and 1783. It was during these periods of volatility that local elites most readily drew on climate theory to try to entice British and northern European migrants to settle permanently in the empire's newest northern colony.[18]

"A Colder and Less Plentifull Country than New England"

Fishermen, merchants, and other itinerants engaged in the Atlantic fisheries had put coastal villages between the Narragansett Bay and the Northumberland Strait in frequent contact since the late sixteenth century. Through the seventeenth century, the expansion of the offshore fisheries centered on George's Bank further strengthened the region's maritime economy. In the meantime, colonists in New England and New York began to regard their northeastern frontiers with the Iroquois, Abenaki, Mi'kmaq, Maliseet, and French with a mixture of fear and envy, eyeing the forts and coastal ports from Montreal to the Gulf of Maine as military targets and commercial opportunities. During and immediately following King William's War (1688 to 1697) they attempted to reduce their vulnerability to a neighboring Catholic empire and its Native American allies while gaining greater access to the interior fur trade and Grand Banks fisheries. But through the turn of the century few English colonists who participated in or supported the raids expressed interest specifically in acquiring Canadian or Acadian lands as green acres for their own use. Samuel Vetch's "Canada Survey'd" was, if not the first, certainly the most forceful articulation of a climatic argument for why the British should deliberately turn to settler colonialism in the region.[19]

Vetch had been one of the council members involved with the Company of Scotland Trading to Africa and the West Indies' brief colonial experiment, the Caledonia settlement at Darien on the Isthmus of Panama. In pursuit of a Scottish Empire in America, the Company hoped to establish a fortified trading post between the Atlantic and Pacific Oceans and eventually a plantation colony. During its short life from 1698 to 1700, the colony suffered from mismanagement and frequent Spanish assaults. Some 2,000 out of 2,500 Scots died, the majority from yellow fever, malaria, dysentery, or exposure. All capital invested in the disastrous venture was lost. Predictably, promoters had proclaimed the local climate to be perfectly balanced, annoyed by "neither pinching Colds, nor scorching Heat"; survivors of the ordeal decided instead that "our settlement in Darien was in a very sickly and unwholesome Climate." More specifically, the tropics were unhealthy for Scots. For "People of a far more Northerly Latitude than Spain is," Central America proved to be "a Place too hot for them." In the summer of 1699, Vetch and other leaders of the Company abandoned Darien to return to Scotland, stopping over in New York City for provisions.[20]

Vetch decided to remain in North America, where he resumed a career of opportunism, narrow escapes, and the pursuit of harebrained schemes with a low rate of success. In New York he instantly forged connections to local elites. In 1700, he married Margaret Livingston, the daughter of Robert Livingston, a fellow Scot who had become a prominent merchant in Albany and New York City, member of the New York Provincial Assembly, and Secretary for Indian Affairs. Extending Livingston's social and commercial links, Vetch began a smuggling operation dealing in furs and liquor among New York, New France, New England, and Acadia, in the course of which he cemented relationships with local merchants and politicians and familiarized himself with the coastal settlements between Long Island Sound and the Gulf of St. Lawrence. In Boston, he ingratiated himself with Francis Nicholson and Joseph Dudley. At the beginning of the War of Spanish Succession Vetch relocated to Boston, where Dudley appointed him to lead a diplomatic mission to Quebec. This appointment also proved convenient for resuming illegal trade with New France, including the sale of arms, which Dudley tacitly allowed. In 1706 Cotton Mather and other powerful opponents of the governor made a public scandal of Vetch's crimes; he was convicted and almost imprisoned. But Dudley helped Vetch flee to London, where the Privy Council overruled the Massachusetts General Court and acquitted Vetch of all charges.[21]

Vetch arrived in London just in time for negotiations over the Act of Union, which, he immediately grasped, opened new commercial and colonial opportunities for Scots in the now–British Empire. Channeling his detailed knowledge of the northeastern North American coast and a conviction that his countrymen would do best to colonize cold climates, Vetch collaborated with Dudley and Nicholson on "Canada Survey'd," several drafts of which they submitted to the Board of Trade between 1706 and 1708. Their plan expanded on previous campaigns by New Englanders against Acadia and New France and recommended two actions: first, a military expedition, and second, an emigration scheme. To "put an end to the troubles upon the whole shore of America," both parts of the plan—conquest and British settler colonialism—were crucial. "As the french grow more numerous" British America would be "sourrounded and hemmed in betwixt them and the sea . . . so that in time, when they are fully peopled, as they project in a great measure to be, after the warr is over, by transporting thither (as Monsieur Rodot, the present Intendant of Canada, told me the french King designed), 20,000 men." To counteract the king's design, the British would mobilize their Iroquois allies in northern New York together with 2,200 colonial troops, sailors, and officers from every colony north of Maryland.

After the conquest of territory between Montreal and Port Royal, they would "very easily remove the French" from this vast region and repopulate it with loyal Protestants, particularly "idle" Scots who would better contribute to British security and prosperity abroad than burdening it at home. "To give the finishing stroke to all," the proposal's authors compared Canada to New England, where "the healthyness of the climate, which with the genius of the people calculate to improve all those advantages, have rendred itt a place of vast trade and buisiness, for besides that the inland country affords great quantityes of all sorts of provisions, horses, cattell, and lumber, fitt to be transported to the West Indies and elsewhere." The northern extension of British colonies—"notwithstanding the coldness of their winters"—would further enlarge the empire's provisioning grounds. These political and climatic calculations rendered "unanswerable arguments for reducing Canada" and making it into a British colony of settlement—the most economical, decisive check to the otherwise inevitable growth of French American population and power.[22]

Reprising the association between cold climate and Britishness, in February 1707 Vetch flattered the queen with a birthday poem dedicated to "Immortal Anne. . .Whose Colder Northern Blast" would "quickly Chill"

Louis XIV's "Sun-shine Plants." The Queen approved Vetch's plan to seize Canada. As it happened, the campaign against New France faltered, but Acadia was an easier target. In 1710 Nicholson and a force of nearly 2,000 men, most of whom were New England troops, attacked the garrison at Port Royal on the Bay of Fundy coast, where the outnumbered Acadians surrendered. Vetch became the first British governor of the province, succeeded soon after by Nicholson. At the end of the war in 1713, New France remained French but the Treaty of Utrecht ceded Nova Scotia to Britain.[23]

To secure the conquest, both Nicholson and Vetch continued to lobby the crown and Lords of Trade to subsidize the settlement of Nova Scotia, reiterating the formula for success spelled out in "Canada Survey'd." Because Nova Scotia's "climate being much more callculate for the British constitutions then the more southern setlements in the West Indies," they had argued in 1710 and 1711, sending 400 to 500 British and Irish Protestant families would be in the "intrest and advantage of the Brittish Empyre in generall." Vetch and Nicholson implored the Queen to encourage British migrants "as she was pleased to do for the Palatines" who had recently been transported to New York. The Palatines were German Protestants, roughly 15,000 of whom were stranded in refugee camps on the outskirts of London. Facing economic distress in the Rhineland, they had responded to British recruiters who promised them free passage and land in New York, Pennsylvania, or the Carolinas. But their arrival in England during the bitterly cold winter of 1708/1709 also coincided with, and in turn fueled political controversy over, passage of the General Naturalization Act. The Board of Trade suspended support for most of the proposed schemes, and the majority of refugees were forced to return home. The main exception was a group of over 2,000 Palatines who settled near Albany to work on lands controlled largely by and to the financial advantage of Vetch's father-in-law, Robert Livingston. But this family connection did little to help Vetch's own bid for the subsidized transportation of Scottish Protestants to Nova Scotia.[24]

Vetch never entirely relinquished the idea of remaking the province into a northern extension of Britain's settler colonies. He continually urged the Board to reconsider "settling the main coast of Nova Scotia with all imaginable speed" because "it is every way framed by nature to make one of the greatest and most flourishing settlements in all America." At the same time, he frankly described the region's hard winters in negotiations with his superiors. The soldiers at what was now known as Annapolis Royal were never given adequate equipment, but only "rotting" uniforms without lining that seemed "rather calculated for the Torrid zone than the Inhabitants of one of the Coldest countrys in the world." They were living "upon the Levell with the slaves in the neighbouring Colony of New England." This was an especially pointed comparison because, he proclaimed, "it would be very hard to confine a parish of men to a garrison in a colder and less plentifull country than New England."[25]

The fractious, shivering soldiers at the chronically undersupplied garrison were the only British permanent residents in Nova Scotia. In 1709, Vetch, Nicholson, and Dudley had also faced political resistance to expelling the Acadians and their Mi'kmaq allies. The new Tory ministry decided against these terms of the original plan, believing that allowing the colonized inhabitants to remain would provide useful local knowledge and manpower. By the time Vetch was reappointed governor in 1715 he and other local officials also saw the value in retaining them. In 1716, he opposed the Board's new intention to remove the Acadians and Mi'kmaq to Île Royale (Cape Breton), where the French had just founded Louisbourg. The Acadians would leave with their crops and black cattle, which had become habituated to the local climate—"the agreeableness of the Soil & Climate to those Creatures being the same as they were bred in which will mightily contribute both to their healthfulness and fruitfulness and could not be in Several Years Expected from those transported from france." The deportation would strengthen French power in the area while emptying Nova Scotia of virtually all its European settlers. "As to what may be the consequence of french removing from Nova Scotia to Cape Br they are evidently these: first there leaving that country intirely destitute of inhabitants there being none but french & Indians."[26]

Successive British governors were equally ineffective at attracting substantial numbers of Protestant settlers. Scots migrated to other American colonies, but none came to Nova Scotia until after the Seven Years' War. During the first half of the eighteenth century, the British and French engaged in a series of contests to consolidate their territorial claims in the region, but British projects to change the demography of the peninsula were half-hearted and inconclusive. Unlike the far more lucrative offshore claims to fisheries in the Gulf of St. Lawrence and Gulf of Maine, land in what is now eastern Canada was never a top priority for the British. Aside from seasonal communities of New England fishers and the British soldiers in Annapolis Royal, Nova Scotia remained Acadian and Mi'kmaw territory. Mi'kmaq outnumbered British inhabitants and the French farmers who cultivated the limited arable land in the Annapolis Valley on the Bay of Fundy coast were the majority settler population. From 1720 to 1744, the Acadian population doubled. Although several thousand troops from Boston proudly attacked the fortress at Louisbourg in 1745, negligible numbers of New Englanders settled land in Nova Scotia as a result. When the British established the new naval base and provincial capital of Halifax in 1749, the Acadian population totaled over 13,000, not including neighboring areas under French control or disputed sovereignty. After the Treaty of Aix-La-Chapelle in 1748, approximately 2,700 Protestant migrants from the German countries, France, the Netherlands, and Switzerland settled in the Annapolis Valley or established new towns on the Atlantic coast. By 1752, these groups, together with Anglicans, Congregationalists, and Presbyterians from Yorkshire, Scotland, and Ireland, constituted a non-Catholic settler population of 4,200. A decade later, this number rose to nearly 9,000 as New Englanders moved to occupy fertile farmlands abandoned after the *grande derangement*, or violent mass eviction, of 14,000 Acadians from the province in 1755 during the Seven Years' War. In the years leading up to the American Revolution, small numbers of immigrants trickled into Nova Scotia, but it remained one of the most marginal colonies in British America, relying heavily on external subsidies and active recruitment.[27]

This recruitment continued to focus specifically on Protestant migrants from Scotland and northern Europe. In the 1770s, a group of Scottish colonists, including fourteen proprietors in Nova Scotia and John Witherspoon, president of the College of New Jersey in Princeton, convinced 200 Highland tenants to exchange their rising leases for free passage to the Northumberland Strait, where the "lands seem to be calculated better for a Scots settlement than most others upon the continent of North America." Those who survived the misfortunes of their transatlantic voyage arrived in mid-September to find unimproved land, sharp weather, and none of the promised material support, just as "A Wellwisher to Old Scotland" who responded directly to the proprietors' land advertisements in the Scottish press had predicted. No price was too low to settle in that "dreary tract of uninhabited and uncultivated wilderness." Anti-emigrationists denounced Witherspoon's false suggestion that "Nova Scotia cannot be more disagreeable to the Scottish, than Canada is to French constitutions."[28]

Undeterred by bad publicity, local elites held fast to the principle of northern acclimatization. Frustrated by the ongoing struggle to maintain a sufficiently large colonial population, Isaac Deschamps suggested the province redouble its efforts to attract Protestant migrants from northern European climates by instituting "a regular scheme." After the Acadian expulsion, he argued, the "scarcity of settlers" was no longer a matter of territorial sovereignty but instead the more basic problem of chronic labor shortage, especially during the planting and harvesting seasons when workers were "most wanting." Deschamps was a Swiss Protestant who came to Nova Scotia in 1749, became a large proprietor in the Annapolis Valley, and served in a variety of provincial offices. He crafted a recruitment policy incorporating headright with climate theory, modeled on what he considered as two exemplars of colonial development: first, the land granting system that had drawn indentured servants to seventeenth-century Virginia and Pennsylvania; and second, the 1753 settlement of Swiss and German Protestants at Lunenburg (previously, the Acadian town Mirligueche) on Nova Scotia's Atlantic coast, where they "left nothing undone to forward their prosperity." According to Deschamps's plan, Nova Scotia's large proprietors would pay for the transportation and subsistence of Protestant migrants who, after serving four-year terms, would receive equipment, raw materials, livestock, and title to fifty acres with a twenty-year terminal quitrent.

Believing that northern European Protestants were more likely to be "Honest industrious people inur'd to Labour," in his proposal he combined "the spirits of the Virginia Company" with that of "Canada Survey'd." Deschamps did not specifically mention Vetch, but their explanations for why land agents should target northern European emigrants were strikingly similar. Deschamps figured that "Foreigners, British, and Irish" Protestants were simply better accustomed to working in the cold outdoors:

> The northern Inhabitants of Europe would prefer this Country to any other in America being nearly the same Climate as their own and dreading any climate warmer, those are the people we should introduce to make this a flourishing country they being inur'd not only to a cold climate but to every other labour this country wants to make it flourish. Whereas people born in a warmer climate will not think themselves happy in a cold one.

Deschamps's argument was highly partial and inconsistent with contemporary evidence. The New England Planters—migrants who colonized Acadian lands in the immediate postexpulsion period—fit Deschamps's description of Protestants accustomed to the difficulties of northern farming. But "nothing of any consequence in husbandry could be expected" from them, Deschamps wrote dismissively. And although he praised "the valuable Improvements left to us by the Expulsion of the French Acadians," he did not consider the remaining Acadians in the province—by his count, 100 families who had eluded deportation, retreated to the peripheries of their former villages, or returned to the area as tenants of British landowners—as legitimate settlers. The common view was that, though these dispossessed French farmers might be permanent residents, they "were not settlers in terms of the grants."[29]

Deschamps proposed this scheme in 1782. Coincidentally, the following year the province received its largest influx of British immigrants: well over 30,000 evacuees from the United States, including 3,000 African American Revolutionary War veterans, former slaves from southern colonies who served with British regiments in exchange for freedom and land grants. As a group, the Loyalists were a welcome answer to the long-standing "desire" of the colony's "administration to increase the population of Nova Scotia." Their arrival had nothing to do with the climate but, as Joseph Fredrick Wallet DesBarres recognized, it was one of the reasons why many of them came to view Nova Scotia as a temporary refuge rather than as a final destination.[30]

DesBarres was a Swiss–British engineer who came to Nova Scotia after the Seven Years' War to complete a hydrographic survey of the coast from New York to the Gulf of Saint Lawrence. While working on drafts of this project, he acquired extensive tracts in the Annapolis Valley and built an estate in Falmouth he modestly named Castle Fredrick. As compensation for his survey *The Atlantic Neptune*, published as a series from 1774 to 1784, the Admiralty rewarded him with political appointments and additional land grants totaling 80,000 acres in Nova Scotia and Cape Breton Island (ceded to Britain in 1763). From 1784 to 1787 DesBarres was the lieutenant governor of Cape Breton; from 1804 to 1812 he was lieutenant governor of Prince Edward Island. Among his ideas for regional economic development was an attempt to engage Loyalist whalers from Martha's Vineyard, Nantucket, and Rhode Island to relocate the base of their fisheries to Cape Breton. The failure of this project made him skeptical about the commitment of the more "substantial kind" of settler from New England—the "wealthy Nantucket and Newport whaling families" who were "unwilling to subject themselves to this harsh prospect" and promptly emigrated. They "sold the lands that were granted to them and meanly skulked into the United States."[31]

Like Nova Scotia's many other absentee landowners, DesBarres preferred living in London. Unlike some of them, he recognized the limits of northern acclimatization and did not pretend his Swiss birth inured him or anyone else to local winters—"little inferior to that of Russia," he wrote. "Many arrived" in the eastern seaports of America "from Europe, who almost invariably found their way from thence to the present American States, from whence none would go to settle Nova Scotia, more than they would go to Siberia." Although Nova Scotia's "climate is remarkably healthy and this province affords as striking instances of longevity as any upon the continent of America," DesBarres knew that the climate had also been a common reason for the departure or displeasure of settlers throughout the century. According to a new Scottish arrival to Prince Edward Island in 1791, acclimatizers had miscalculated the northern temper. This man implored his former landlord in the Outer Hebrides to prevent others from settling in the northern colonies: "It would be best to stay at home for their Constitution will not answer to the climate here."[32]

The Poor Man's Dung

For those who decided to stay, a comfortable existence was possible in the region, but it required coming to terms with the possibilities and constraints of a cold, northern environment, including following the practices of other northern farmers. An Irish emigrant and land agent responsible for settling one of the most barren tracts in Prince Edward Island thought DesBarres's comparison of the region to Russia might be self-deprecating but it was apt. As he told DesBarres, "I was very much charmed by the Island, I determined to spend my days here: and I have invariably endeavored to render my asylum valuable and agreeable by introducing system." He proposed the establishment of a local agricultural society following "the Oeconomic Society of Russia both in the name and the design." The similarity between Russia and northeastern North America's climates struck many observers. Massachusetts Loyalist Mather Byles, Jr., who resettled permanently in New Brunswick, called the Maritimes an "American Siberia." Botanists in Halifax were impressed by Catherine the Great's flourishing gardens in St. Petersburg and Arkhangelsk, "where, under a wintry sky, the magnificent plants & luxurious fruits of the tropics find a home." In the District of Maine, Benjamin Vaughan overwintered "trans-Atlantic vines" underground as he had heard was done by grape growers in "Astrachan & other places in Siberia."[33]

These were instances of local elites' larger argument that northeastern North America was located in a cold temperate climate zone encompassing Russia, the British Isles, Scandinavia, northern Europe, and the Eastern Seaboard north of Maryland. Timothy Dwight only half-jokingly compared his fondness for New England to that of "the Laplander" who "believes the frosty region around him to have been the seat of Paradise" and "an Icelander" who "can find a comfortable life no where, but in the dreary island in which he was born." Naturalists in the region were particularly interested in Carolus

Linnaeus's garden and menagerie in Uppsala—the most ambitious northern acclimatization effort of the eighteenth century, which was assembled with the help of a global network of collectors. Linnaeus's disciple Pehr Kalm gathered North American specimens from Pennsylvania to Quebec but he did not stop in New England or Nova Scotia. Kalm assumed from other travelers' reports that the winters in the Northeast could be worse than in Philadelphia, New York, or England, but "full as severe as our Swedish winters." By this act of omission, he unwittingly reinforced the idea of greater New England's inclusion in a broad northern climate zone. Familiar with Kalm's comments, when the naturalist William Peck traveled to Scandinavia, he also noted the physical similarities between the regions. The steep northern coast of Jutland in Denmark reminded him of the "White Mountains in New Hampshire," because it was "covered like them with dark brown moss." And as he wrote to his sister: "I must tell you . . . that next to my own country I love Sweden, because it so resembles it." Sweden was especially admired in Nova Scotia, where improvers marveled at how Sweden,

> a country which we would think scarcely habitable, or worth cultivation, abounds not only in the necessaries, but in all the conveniences and comforts of life. Sweden is one of the most northern and barren countries in Europe. Stockholm, the capital, is nearly in the latitude of 60 degrees—almost one thousand miles to the north of Halifax. The whole kingdom is overspread with rocky mountains and lakes, having little land capable of culture, and is subject to all severities of so high a Latitude. But Sweden has been fortunate in producing a number of eminent men, who made great improvements in Natural History—particularly in agriculture."[34]

On the other hand, Linnaeus's public failures with acclimatizing tropical plants in Swedish greenhouses suggested that there were limits to this technique. Species accustomed to one climate were not infinitely adaptable to all places. The seeds of warmer-climate species would not develop with limited heat and daylight or survive in soil hardened by frost. The key was to import seed stock and agricultural practices from similar climates. Samuel Deane, author of the agricultural dictionary *The New-England Farmer*, agreed with the Scottish improver Adam Dickson, who mocked anyone "so foolish to suppose, that all kinds of plants can be cultivated with equal success in all climates." Deane drew attention to successful experiments with particular crop rotations recommended by European improvers, but cautioned farmers that specific advice could not always be generalized, as "soils differ so greatly, even in fields which lie contiguous, that the course of crops which is suitable for one, would be unsuitable for another." Instead, he drew heavily on the voluminous published works of Scottish authors, such as the accounts of surveys and tours of the Highlands written by Pennant, Sinclair, and Johnson and Boswell, reproducing them for American readers through thinly disguised or outright plagiarism. Much of Deane's supposedly homegrown advice was in fact transplanted directly from Scottish and Irish agricultural literature.[35]

Visitors from the other side of the Atlantic also saw the resemblance of northern climates, species, and agricultural practices. Two speculators from York judged the weather in Nova Scotia "pretty near that of England"; a Halifax farmer declared "England is like this a Cold Northern Climate." The Nova Scotia Society for Promoting Agriculture suggested there was "the same difference between our spring and that of New York," which was something like "spring in Middlesex and that of Yorkshire, in England." Despite the fact that Yorkshire was "in Latitude 52 degrees," its farmers were prosperous and so, it followed, "that the same quantity of land, acre for acre, in Nova Scotia, will maintain as many people, yield as much corn, as in New York, New Jersey, Pennsylvania, or any of the old Colonies."[36]

Some writers interested in acclimatization distinguished among northern, northeastern, and southern areas of the region, but most minimized the climatic differences except insofar as farmers made slight adjustments to their schedule of seasonal activities. Paul Mascarene was certain that though Nova Scotia's "climate is cold and very Variable even in ye Southernmost part of this Country, and is subject to long and severe Winters, the soil, notwithstanding this, may be easily made to produce all the supplys of life for the Inhabitants." A 1748 account that included all of New England and Nova Scotia, Cape Breton, and Newfoundland in its geographical "description of the northern colonies" was more emphatic. It claimed that "there is very little Difference in the Temperature of the Air, in the several Parts of New-England, so its several Products and aptness for different Improvements, vary but in a few Particulars," namely, the "southernmost" areas were best suited for growing grain, whereas lands farther north were better for forestry, fisheries, and raising livestock. Although northern New England was decidedly "cold in winter," the summers were sufficiently hot to grow "Indian corn & melons in the open air." A Rhode Island improver disputed a "valuable treatise on husbandry," which claimed that northern American soils froze too hard to bear winter cover crops like lucerne (alfalfa): "It may be so in the province of Maine where the author resides; but it is certain that on Rhode-Island, the lucerne bears the winter equally well with clover and other cultivated grasses . . . It is probable it may be cultivated in any part of New-England, even in the coldest." Halifax agronomists remarked "our spring is indeed later than in countries that lie farther south; but countries which are north of us, and whose spring is later than ours, abound in provisions.[37]

Timothy Ruggles and Edward Winslow, prominent Loyalists whose ancestors had been among the first colonists in Massachusetts and Plymouth, were certain that they could reestablish their lives in Nova Scotia and New Brunswick with all the physical comforts they had enjoyed in southern New England. Ruggles acquired 10,000 acres on former Acadian lands in Wilmot, a township in the fertile Annapolis Valley on the coast of the Bay of Fundy. In 1783 he described his flourishing walnut and apple orchards to Winslow. At first he had been "apprehensive from the scantiness of heat & much wet weather" that maize would "not be a proper grain for this climate," but these fears were allayed by an impressive harvest. "Upon the whole," Ruggles concluded about Nova Scotia, "I think the climate good & the soil capable of becoming the granary of any part of the continent to the Eastward of New York."[38]

Settlers who prospered in the region stressed that winters, freezing temperatures, and large quantities of snow were undeniably difficult to bear but also offered some benefits. Although frozen water slowed long-distance transportation, it allowed for quicker river and pond crossings. Some vegetables, like parsnips, were better preserved underground in northern New England's frozen soil than in cellars. Hard frost also suppressed agricultural diseases. Many observers thought that the cold explained why the Hessian fly—a midge that attacked American and British barley and wheat fields from the 1770s through the turn of the century—did not reach northern New England and Nova Scotia until the 1790s.[39] Ministers asked parishioners to contemplate the providence of winter's effects.

> Long Experience has taught us that by Means of a large Coat of Snow upon the Earth, (by which it is preserved from extreme freezing, or the Frost taken out by slow Degrees, by the comparatively warm Coverlet of the Snow) our Fields are less torn to Pieces by Rains: are left softer, lighter, and as it were, enrich'd with a coat of Dung; for the better Growth of Vegetables, or the Fruits of our Fields in the Spring and Summer. Snows are the Treasure of God. From his Store-House on high, where Mortals cannot approach, they descend.

A common proverb was that snow was both "the poor Man's Dung" and "the rich Man's too, in these northern Climates." Naturalists offered a more scientific explanation of this saying: snow insulated, watered, and enriched the soil. While solid, it acted as a protective layer that preserved the ground's "inner heat"; as it melted, it released water that saturated topsoil with fertilizing minerals. Overall, "the advantage derived to the earth from the quantity & duration of the snow is everywhere apparent," explained Samuel Williams.[40]

Despite these local advantages, the agrarian economy was "the most dismal" compared with the rising wealth of other British American regions. Immigrants with smaller means who settled in the Northeast had higher prospects of securing modest landownership and establishing independent households. With the exception of Connecticut River Valley tobacco farmers, however, other farmers found that much of the region was unsuitable for growing staple crops on any considerable scale, and no single crop dominated the region's rural landscape or exports. The Northeast's connection to the maritime economy in the seventeenth and eighteenth centuries was based in port cities like Newport, Boston, Portsmouth, and Halifax, which, besides exporting timber harvested from northern New England's forests and provisioning livestock to Newfoundland and the West Indies, relied mainly on shipping and the cod and whale fisheries. Most of the region's farms produced for subsistence or local markets and were more reliant on the work of family members or seasonal laborers rather than on that of servants or slaves. The modest scale of operations on northern farms was the result of "local circumstances and hard necessity." Benjamin Waterhouse remarked that the landscape was "different at the southward" because the climate and soil allowed planters to raise "a few articles to a great extent" and plainly "exceed any in the Northern for expensive agriculture."[41]

What northern planters lacked in profit potential, they made up for in the variety of crops, fruits, and livestock of their mixed husbandry farms. Rather than raising the same staples as other northern farmers in the British Isles, some argued that northern colonists should further diversify by acclimatizing specialty crops, especially luxuries associated with Mediterranean or otherwise warmer climates, like saffron, silk, and wine grapes. Capturing consumer demand for Asian and Mediterranean products like silk and wine intermittently figured in the crown's early seventeenth-century encouragements to Virginians to plant mulberry trees, projectors' utopian visions for a horticultural colony in early Georgia, and in later eighteenth-century southern improvers' proposals to encourage cottage industries among small farmers, especially in the Lower South. The earliest to envision the northern colonies as rivals to winemakers in Bordeaux and Burgundy was the Earl of Bellomont, appointed governor of New York, Massachusetts, and New Hampshire from 1695 to 1701. He knew that French courts had banned winemaking in New France, but England had no constituency to oppose colonial vineyards. In 1700, he also recommended that the Board of Trade encourage mulberry plantations and salt works in New England. The only obstacle to realizing these projects would be finding a low-wage labor force in a region of smallholders. "The Indians," Bellomont complained "are so proud and lazy, that 'tis to be fear'd they will not be prevail'd with to work."[42]

After the Seven Years' War, the Society for the Encouragement of Arts, Manufactures, and Commerce in London offered premiums for planting mulberry trees in New England, hoping to promote sericulture (or silkworm husbandry) as a strategy of regional economic diversification. Jared Eliot took up their campaign in Connecticut. That colony needed to specialize its agricultural production in a way so as to avoid challenging British sheep breeders and wool manufacturers; merchants based in the fisheries of Newfoundland, Rhode Island, or Massachusetts; and southern planters for cash crops. For Connecticut landowners, Eliot argued, planting mulberry trees would answer mercantilists' fears while also providing shade, firewood, and soils that would retain more moisture from the greater accumulation of dew. In 1769, Nathaniel Ames promised cash prizes, ranging from $100 to $10 to landowners in southern New England who would raise the most mulberry trees. Noting that "raw Silk is plentifully raised in much more Northern Climates than this," Ames preempted objections that silkworms preferred "hot climates." Although it was true that the worms would develop on trees in the open air in "hot countries," the main adjustment northern growers would have to make would be to nurse cocoons in "a warm room," where they could feed on plucked mulberry leaves.[43]

Metropolitan initiatives to promote silk production in New England continued after American independence. English naturalist John Coakely Lettsom anonymously financed prizes administered through the Massachusetts Society for Promoting Agriculture for planting white mulberry trees in the District of Maine, where he knew that apple orchards thrived. He wanted to encourage "those who begin settlements in new plantations" to invest in northern sericulture and pointed to the success of Prussian silk exports, a young industry that was the direct result of pressure from that country's agricultural society. Even without the encouragement of foreign patrons, however, northerners had resumed planting mulberry trees. In 1790 the *Columbian Centinal* reported that sixty families in New Haven had raised 442,000 silkworms; in Massachusetts, local elites distributed white mulberry seeds "sufficient to produce Eighty Thousand trees." At Harvard, Benjamin Waterhouse included a unit on the silkworm as part of his natural history lectures, in which he first described the worm's "natural progress in the open air of the warm and serene climates of some parts of Asia," before proceeding to explain the artificial methods of raising silkworms in "the rough and unsteady" climate of northeastern North America. Despite these efforts, sericulture remained a negligible sector of early New England's economy as it did in Sweden and Scotland.[44]

In Nova Scotia, regional diversification took a backseat to more basic concerns. The scarcity of labor in the province encouraged officials to focus instead on import substitution of staple commodities, especially for addressing Nova Scotia's dependency on subsidies. As one contributor to the *Halifax Gazette*, who used the pen name Agricola, remarked in 1752:

> The spirit now starting among us to clear and cultivate the lands will I hope be cherish'd: it is certainly of lasting importance to us. We can now behold 'tis true a large and pompous town, but, until we can see the plains and country round about it covered with grass or grain new difficulties will every day arise and the expensive method of our present supply of provisions will reduce the inhabitants (many of them) to a state of insolvency.

After the American Revolution, fears of a downward spiral in the agricultural economy if the region continued to lose population took on more urgency. The British Board of Agriculture was keen to promote hemp, flax, wheat, and potato production alongside the fisheries and timber trade in Nova Scotia because British merchants and the navy were cut off from provisions and raw materials from the United States to supply Caribbean plantations. If these were the only staple crops that had a reasonable chance of success in the region they would also demonstrate that Nova Scotia was "a valuable appendage to the parent state, and a sure source of permanent supplies to our West India islands." West Indian planters, for their part, ridiculed the notion that the "shriveled barley, oats and rye" produced under the "transient gleam of sunshine" of northern summers would be "excellent substitutes for the flour of New York and the rice of Carolina." Consequently, in every newspaper published in the province in 1789, the Nova Scotia Society for Promoting Agriculture explained its motivation to promote import substitution, implicitly addressing the concerns raised by skeptics about whether the local climate would prevent its success.[45]

But to grow northern crops on any reasonable commercial scale, the colony first had to "procure inhabitants." As a result, one of the major goals of the agricultural society was to encourage immigration. Lieutenant Governor John Parr, who was also the first president of the Society, was determined to convince outsiders that coming to Nova Scotia would not represent immigration and displacement but instead relocation from one northern realm to another. It was essentially the same as the northern United States or northern Europe in terms of environmental conditions, he submitted, but superior in every other respect: accessibility of land, low cost of living, and greater personal liberties. "In the present confusions of Europe, there are thousands who would be happy to take sanctuary" in Nova Scotia, "fertile and salubrious; with mild laws, a settled Government, no taxes to be paid, full liberty of Conscience, with plenty of land that wants cultivators, besides many other natural advantages; and all this is under the protection of Great-Britain." Like Vetch and other governing elites had done before him, Parr declared that "if only migrants were sufficiently acquainted with the state of this Province, it would soon turn the tide of emigration to Nova Scotia."[46]

Observing the immigration history of colonial Nova Scotia, in 1790 the Scot Andrew Brown struggled to understand how any settlement scheme could be successful "in this hyperborean region." Brown was a Presbyterian minister educated at Glasgow and Edinburgh, where he met his mentor William Robertson. In 1787 Robertson sent him to lead a congregation in Halifax, where Brown remained until 1795. Shortly after arriving, he began collecting materials and recording his interviews with local elites like DesBarres and Deschamps for writing a history of the province. Encouraged by Robertson and Jeremy Belknap, Brown completed three chapters that treated the Acadian peninsula as an extension of Massachusetts, covering the foundations of Plymouth Plantation to Vetch's appointment as Nova Scotia's first British governor.

If it were true that "the heats of summer and the frosts of Winter were equally fatal" for new immigrants throughout North America, then why, Brown wondered, had Europeans colonized the region at all? This question loomed in the background of two other "important questions" Brown wanted to "enable every reader" of his manuscript *History of the Province of Nova Scotia* to answer. First, what would be the "effects of emigration on the life and character of the individual" who relocated from Britain to British North America? Second, how could prospective migrants compare their current situation in Britain with "the respective advantages of a settlement" in the colonies? He rejected Vetch's simplistic idea that the province would naturally appeal to Britons: When Brown "contemplated the strange landskip" of the province, he "sought in vain for objects of resemblance" to Scotland. *History of the Province of Nova Scotia* bristles with bitter judgment of colonists' militant, unrelenting assault on Native Americans and French Acadians. But to Brown's mind the bigger shame was that the establishment of a flourishing colony did not follow the destruction of these societies. His account ends with a condemnation of the province's many "projectors and patrons," whose attempts to attract British settlers "terminated in disappointment and ruin." As Brown recognized, the first governor of Nova Scotia's most lasting achievement was the invention of a tenacious problem: how to address persistent concerns about the dangers of the local climate so as to remake the peninsula into a British colony of "useful and loyal settlers."[47]

4

Jamaicans In and Out of Nova Scotia

The shudd'ring tenant of the frigid zone
Boldly proclaims that happiest spot his own.
OLIVER GOLDSMITH, "The traveller, or a prospect of society" (1764)

DURING THE LAST four years of the eighteenth century, over 550 deportees from the Maroon community of Trelawny Town, Jamaica, lived in limbo in Nova Scotia. The lieutenant governor of Nova Scotia, Sir John Wentworth, was eager to establish them as permanent residents but strong opposition finally prevented it. The Maroons and concerned politicians in London complained that the northern climate was too cold for black people, who required the hot tropical climates of the Caribbean or Africa to thrive. Although the Trelawny Maroons arrived in summer, both they and their advocates in the British Parliament worried about the approaching winter, arguing that the inhumanity of their expulsion from Jamaica would be compounded by their inability to withstand Nova Scotia's harsh climate. They believed that the province's long winters were "so bad—so cold" that the Maroons would surely "die" if they stayed.[1]

To Wentworth's regret, such declarations reinforced what he and other local elites maintained were false notions about the regional climate. The temporary settlement of the Jamaican Maroons in Nova Scotia from July 1796 to their departure for Sierra Leone in August 1800 was the most extraordinary episode in the long history of attempts to prove to outsiders that the region's climate was temperate. This chapter examines the brief history of the Trelawny Maroons in Nova Scotia in terms of long-standing debates about human difference and the relative habitability of northern climates, focusing especially on how Wentworth and other local elites addressed the transatlantic controversy over settling the Maroons and other black migrants in northeastern North America. By inventively combining several strands of contemporary thinking about climate, race, and migration, Wentworth tried to convince the Maroons and their advocates abroad that they should settle permanently in Nova Scotia. His primary tactic was strenuous denial that the province's climate was dominated by extremely cold, interminable winters. He dismissed nonresidents' fears about the degenerative or even fatal effects of the province's allegedly severe winters as misconceptions.

Like those of his predecessors, Wentworth's public remarks about the northern climate were motivated by the exigencies of settler colonialism in the relatively undesirable corner of empire for which he was responsible. However, unlike his predecessors, he argued that the region was located squarely inside the temperate zone. In this instance, Wentworth's insistence that Nova Scotia's climate was temperate enough for black settlers dovetailed in unusual ways with British abolitionists' mounting arguments for reforming the empire according to climatic theories of racial geography. On the surface, promoting black settlement in wintry Nova Scotia might seem like an anomalous or even visionary attempt at racial integration in an empire structured by the logic of racial slavery, a logic that extended to abolitionism itself. Leading antislavery advocate William Wilberforce and his allies endorsed climatic theories of racial geography. But Wentworth was not a dedicated abolitionist. Although he may have genuinely believed that Nova Scotia's climate was sufficiently temperate for everybody, he was not crusading on behalf of black Britons or slaves in the name of an anachronistic ideal. Instead, his imaginative scheme was an attempt to advance another crucial but much more prosaic goal, one that had continually eluded the British in Nova Scotia: increasing the settlement, labor pool, and productivity of the colony. The key elements of his ideas in regard to the Maroons were comparable to numerous other seemingly bizarre immigration schemes for populating new, marginal, or unstable territories such as Florida, Botany Bay (Australia), and Sierra Leone or for supplying laborers to plantation colonies, such as proposals presented to the Board of Trade in 1802 and 1806 to transport thousands of Chinese settlers to Trinidad and Jamaica. The majority of such relocation projects either miscarried or never came to fruition, but it is mainly with hindsight that they seem to have been predictably untenable.[2]

If the peculiarity of Wentworth's settlement scheme diminishes when it is placed alongside similar attempts to advance settler colonialism in other parts of the British Empire, his attempts to persuade the Maroons and others that Nova Scotia was temperate reveals a largely unexplored aspect of transatlantic debates about climate and race. Although most historians of environment and empire have studied the difficulty of whites in adjusting to the extreme heat and humidity of tropical colonial environments, this chapter analyzes the perceived health hazards for black settlers in cold northern climates from the early modern period through the late eighteenth century.[3]

Ultimately, Wentworth's policy rhetoric about the local climate proved ineffective in practice. In part, it was because his climatic arguments for permanently settling the Maroons in British North America deviated from increasingly prevailing beliefs about the fixed geographical relationships between climate and race, specifically the assumption that black people could not endure cold weather. It was perhaps a particularly shaky proposition in the 1790s, a moment of rising humanitarian activism by powerful members of Parliament, among whom were the abolitionist founders of the Sierra Leone Company, with their competing interests in settling British West Africa. Most outsiders were unpersuaded by Wentworth's attestations about Nova Scotia's mild climate. Instead, northeastern North American winters continued to be widely regarded as intemperate, too cold for anyone's lasting comfort. In this period, as before, most migrants who had the choice to settle elsewhere usually opted for warmer places.

Jamaica to the Maroons

In March 1796, after nine months of ineffectual fighting, British colonists in Jamaica deployed Cuban hunting dogs and 3,000 soldiers to force the surrender of Maroon insurgents based in Trelawny Town in the tropical highland forests southeast of Montego Bay. The Trelawnys were the largest of five independent enclaves of escaped slaves, or maroons, in Jamaica. From 1655, when the sugar island was transferred from Spanish to English rule, these groups resisted incorporation into and lived apart from the colony's plantation slavery complex. Jamaican Governor Edward Trelawny formally recognized their autonomy in 1739 at the end of the First Maroon War and Governor Alexander Lindsay, sixth earl of Balcarres, curtailed it in 1796 at the end of the Second Maroon War, a spillover from the ongoing French and Haitian Revolutions. In spring 1795, the recently appointed Balcarres was unnerved by Trelawny violence and threats to inspire a general slave insurrection. The Trelawnys were likely reacting to highly localized incidents—namely, that two of their men had been publicly humiliated for the crime of pig stealing—and there is little evidence that they were preparing a general uprising.[4] But in the context of regional unrest and successful organized revolt among slaves, maroons, and free blacks in neighboring colonies, Balcarres perceived them to be in a state of rebellion and declared what he later called a "defensive war" to suppress it. By late winter 1796, General George Walpole, who led the attack, negotiated a peace treaty contingent on the Trelawny Maroons' agreement to relocate to the plains closer to Montego Bay and to forfeit runaway slaves who had joined them during the conflict. In return, the colonial government assured them that they could remain in Jamaica. But in the event, Balcarres disregarded the terms of the truce and resolved to deport the entire population of Trelawny Town to form "some settlement" in another part of the British Empire.[5]

Balcarres's problem was where to send them. The Jamaican Assembly considered the Bahamas, but quickly rejected it and any other plantation colonies in the region. During the era of Atlantic wars and real or rumored slave uprisings throughout the Caribbean, colonial governments with large slave populations were unwilling to accept an influx of free blacks with a history of militancy and insurrection. The Assembly stipulated that the Maroons "ought to be sent to a country in which they will be free"—either some part of British North America or to Sierra Leone, a British settler colony recently established for free blacks.[6] William Dawes Quarrell, who took part in the assault on the Trelawny Maroons and was chosen to supervise their deportation, proposed "disposing" of them in Upper Canada (present-day Ontario) because it was "as remote from the ocean as possible" and would crush their "every idea of ever seeing Jamaica again."[7] The Assembly liked Quarrell's proposal but never officially approved it. While Balcarres and the Duke of Portland, secretary of the Home Office, came to a consensus about a final destination, Quarrell and his deputy Alexander Ochterlony brought 568 Maroon men, women, and children to the port of Halifax, which was supposed to serve as a temporary stopover. When they arrived in late July, Wentworth and Prince Edward, Duke of Kent—who was stationed in Halifax from 1795 to 1799 as commander of the forces in Nova Scotia and New Brunswick—were impressed with the Maroons' "perfectly quiet, orderly and peaceable" behavior. The deportees struck them as quite unlike that of "other classes of negroes." Wentworth proposed to Portland, one of his long-standing London patrons, that they remain in the province.[8]

At first Wentworth hired 150 men at a daily rate of ninepence to complete construction on the city's earthwork fortifications. He also intended to engage them alongside colonial riflemen and Mi'kmaw conscripts to guard the harbor against potential French attack. Since the departure of a large portion of the loyalist refugees who swelled Nova Scotia's population after the American Revolution, the province had struggled to attract or retain permanent settlers. These departures from the province included over 1,000 African American veterans and their families—former American slaves who served the British during the Revolutionary War in exchange for freedom and land grants in Nova Scotia. These disbanded soldiers, the so-called Black Loyalists, arrived to Nova Scotia in 1783, and many of them left it to colonize British Sierra Leone in 1792. Wentworth lamented that the emigration of a large group of Black Loyalists to Africa had "diminished useful laborers," as well as "the supplies of small provisions and vegetables they brought to market" (the latter problem was intensified by a severe drought and subsequent fires in the summer and fall of 1792). In 1796 the labor shortage continued. Permanently settling and employing the Maroons would therefore have immediate benefits to the province. By replacing white soldiers with Maroons, Wentworth would reduce the cost of maintaining the militia as well as release farmers "to assist in the Harvests and other Civil Occupations, which are now much distressed for want of hands." It would also benefit the empire. In any colonial territory with dense forests, he argued, the Maroons would be "very dangerous," but in Nova Scotia's stunted pine barrens and "difficult rocky country" on the Atlantic coast, they could do "no material harm."[9]

In the meantime, Walpole—disgraced and infuriated by the nullification of the peace treaty he had brokered in Jamaica—resigned his commission and returned to England in autumn 1796, entering the House of Commons in support of the Opposition. In April and May 1797 during debates over the Abolition Bill, Walpole pressed for an inquiry into Balcarres's actions as well as for alternative exiles for the Maroons. Violation of the peace agreement and banishment of "a whole people" was a gross violation of rights, he argued, but sending them from the "hottest climate under the torrid zone to the coldest region in North America" was vicious. If they would not be allowed to return to Jamaica, they should at least be transported to "a more temperate climate"—one in which "they could live."[10]

Wentworth understood the situation differently. He was prepared to welcome nearly any sort of immigrant to bolster the settler population, and he strongly objected to the negative characterizations of the climate's allegedly harmful effect on the Maroons, maintaining that it was temperate enough to be healthy for all people, white and black alike. Since arriving in Halifax Quarrell had been suffering from a "gouty bilious complaint," but Wentworth predicted he would soon "yield to the salutary influence of our climate, which seldom fails to recover such cases." The Maroons, on the other hand, were presently in "perfect health" and would remain that way through the winter as long as they were "well fed, warmly cloathed, and comfortably lodged." To insulate them from the worst part of the season, he would outfit them in woolens "of a quality fit for laborers" and supply them with wood for building "warm convenient houses." A physician had accompanied them from Jamaica and would remain to serve them in Nova Scotia. With these provisions, the winter was immaterial. Wentworth was confident "of this fact, eno' to establish it and leave me not affear of success in this case."

More important was gaining the Maroons' confidence. From his "long experience with Negroes & Savages," Wentworth believed that the key to securing their allegiance was to persuade them that they were valued members of the colonial community. If "equally protected and encouraged as other of His Majesty's subjects," the Maroons would prove "useful to us and happy in civilized employments."[11] To foster their loyalty, he specified that the men's uniforms should have large white metal buttons decorated with elements of the Jamaican flag: an alligator holding a sheaf of wheat and an olive branch, emblazoned with the counterfactual slogan "Jamaica to the Maroons—1796." The Home Office, he added, should anticipate future orders of these special buttons, as "more will be annually wanted." Portland and Balcarres agreed to Wentworth's plan and promised him regular subsidies through at least 1798 to cover the expense of establishing the settlement. In the House of Commons, Walpole's opponents referred to Wentworth's correspondence with Portland, which they believed threw sufficient doubt on winter's peril to the Maroons. On May 1, 1797, Walpole's motion was defeated.[12]

In and Out of Nova Scotia

Like that of many nonliterate, marginalized, or deliberately elusive communities, the Maroons' experience of Nova Scotia has been preserved only indirectly through Wentworth, Walpole, and various other officials' reports, correspondence, and petitions on the group's behalf. These sources reveal more about colonial and imperial attitudes about climate and race than about the Maroons' thoughts and actions. According to these accounts, initially the Maroons were reasonably content. Wentworth claimed that they repeatedly told him they were happier there "than ever they were in Jamaica." But as the weather inevitably turned colder, the Maroons became uncomfortable and dissatisfied. They had been promised grants of over 3,000 acres of contiguous farmland, building materials for permanent housing, and warm clothing. Instead, Wentworth placed the hundreds of deportees in shabby outbuildings surrounding a two-story house (called Maroon Hall but reserved for Commissioner Alexander Ochterlony's living quarters) in Preston, a village on the outskirts of Halifax. In an unfortunate coincidence, the woolens did not arrive before the onset of winter, which proved to be "the longest and most severe winter known since the settlement of the province." By spring 1797, Maroon headmen, or captains, were regularly complaining about their circumstances. An early frost destroyed the potatoes they planted in the fall, and the deep freeze of winter spoiled the provisions stored in their cellars. Even if the crop had succeeded or the stored potatoes had remained edible, the Maroons would have been unsatisfied because they craved tropical fruits and vegetables like yams, bananas, and cayenne pepper. They also wanted to hunt for wild hogs. These foods were not only basic to their diet but also prestige items that captains expected as prerogatives. They demanded more rum, coffee, sugar, and cocoa, but Wentworth considered these to be "indulgences" and refused to supply them in sufficient quantities. In the House of Commons in February 1798, Walpole read aloud a petition sent from the Maroons the preceding August, communicating their desperation to leave. It stated that "they were ready to yield themselves sacrifices to the laws, if they had offended them, and were desirous that such of them as were taken with arms in their hands," in Jamaica "might be shot, if their wives and children should be permitted to remove to a climate where they could live."[13]

Wentworth knew that the northern climate, especially during a bad winter, could be "uncomfortable" for the uninitiated. He acknowledged that the Trelawny Maroons were shocked by their first experience of Nova Scotia's winter. They were "naturally alarmed at the continuance of cold and snow & express apprehensions that they cannot maintain themselves by their labor in a cold country." But from his personal experience, he was certain that the climate was not the real issue. When he and his wife Frances moved to Halifax, she too was troubled by the cold winters and the brief but "dreadful hot" summers. Through the 1780s and 1790s, Frances described her "unpleasant situation" in Nova Scotia in letters to her friends in London. In early October 1784 she complained to her confidante Lady Fitzwilliam that "this climate to those who like cold winds will be very grateful. I am now freezing by the fire which we have begun to make a fortnight ago and which will continue necessary until July." But by 1799, she had grown accustomed to her life in Nova Scotia and, in good moods, even appreciated the winters. Writing to Fitzwilliam in late December, she wished for "a Bridge between us [that] could allow you to drive to this place at this very moment. The Ground is covered with snow, the sun shining enough to put ones eyes out, the sky a summer sky—no wind, mild, and the water melting and dropping from the eves of the houses."[14]

Frances Wentworth's changed perception reflected a common belief expressed since the seventeenth century: Many writers maintained that, with the right precautions and temperate habits, the region was fit for new settlers. William Wood was aware that "it may be objected that it is too cold a country for our English men, who have been accustomed to a warmer climate, to which it may be answered," that the dense forests provided wood for building and enough fuel to last the brief winters that "lasteth but for two months or ten weeks." Even the "sharpest air" of the region's winters could be relieved with a bit of exercise, another early account reassured newcomers from England. If anything, keeping cool in summer was the greater challenge, especially for "such who are not as yet well naturalized and inured to the climate," wrote William Hubbard in 1680. As an obituary for a Boston man who died at ninety-three (his twin brother lived to seventy) explained, "the people born in New England, (notwithstanding what has been said of the Inclemency of the Climate) may hope with Temperance and Care to live as long as they do in Europe." A temperate lifestyle would include drinking "young, soft" (non-alcoholic) cider and eating locally grown "Cranberries, Barberries, Currents," which were "better adapted to the health of the Inhabitants than any other fruit," advised John Adams. These were all plentiful in New England as "a kind Providence ordered the productions of the Earth to grow in a manner adapted to the circumstances of the Clymate." On the other hand was the "intemperance" of young men and women who followed the fashions of "inhabitants of milder and more equable climates" by wearing "few and thin garments in the severe seasons." Timothy Dwight criticized "a young lady, dressed *a la Grecque* in a New England winter [who] violates alike good sense, correct taste, sound morals, and the duty of self-preservation." If the Maroons exercised self-discipline, they too would be appropriately equipped for Nova Scotia's winter conditions.[15]

Believing the Maroons would eventually acclimate, the lieutenant governor decided that their complaints about local conditions were merely petty grievances, symptomatic of their recalcitrance, indiscipline, and tendency to "create little troubles to themselves." He also faulted the Maroons for their profligacy. They ate heavy food and did not exercise. They wasted cordwood. Wentworth observed that they ran out of supplies before winter's end because they overheated their houses. At the same time they had to be convinced to dress appropriately to the weather: Although they kept their houses "hotter than fever heat," outdoors they refused to wear any shoes or stockings. After the first winter, several lost toes and fingers to frostbite and a few dozen died from exposure or disease—the original 568 deportees had decreased to

532. Once inured to the winters, however, Wentworth believed that moral improvement would gradually undermine these "natural capricious and cunning" tendencies by substituting "other objects of pursuit & ambition." But such fundamental reform took time. Permanent settlement in Nova Scotia was therefore not only justifiable but also necessary to transforming them into industrious, reliable subjects.[16]

To guarantee "the progress of their improvement," he planned to reorganize the gender division of labor in the community, remaking the men into small farmers who would be "content with a good farm and land to cultivate." He promised to place families "on little farms, with the means of culture & raising stock." In the short term, war with France made it more expedient and "much cheaper" to employ the men in military service, but "when peace is restored" he would beat their swords into plowshares. During his governorship, Wentworth was also president of the Nova Scotia Society for Promoting Agriculture. He made Quarrell and Ochterlony members of the society and gave them wheat, barley, and rye seed to distribute among the Maroons. Maroon men sometimes agreed to work as day laborers, clearing land and cutting wood for white settlers, but it is unknown whether they planted the seed grain. In the meantime, Maroon women continued to be in charge of the community's agricultural work, just as they had been in Jamaica. Wentworth reported that women and children regularly appeared at the Halifax farmers' market to sell berries, eggs, poultry, and baskets "by which they gain great profit in aid of their support."[17]

Besides urging Maroon males to become yeomen, Wentworth wanted to Christianize the entire community. For this endeavor, he hired two ministers: Theophilus Chamberlain and Benjamin Gerrish Gray. Chamberlain knew Wentworth from New Hampshire, where he had studied to be a Congregational minister at Dartmouth under the college's founder Eleazar Wheelock. After missionizing among the Iroquois in New York in the mid-1760s, he relocated to southwestern Connecticut and converted to the Scot Robert Sandeman's apolitical, heterodox sect, which had a tiny but fervent following of about ten families in Danbury and New Haven. Facing harassment for their pacifism during the American Revolution, in 1783 he fled with other Sandemanians to Nova Scotia, where from 1785 to 1792 Chamberlain oversaw land distribution in Preston to American refugees, including Black Loyalists. Gray was ordained as an Anglican minister in England in 1796 and went to Nova Scotia the following year as a missionary for the Society for the Propagation of the Gospel in Foreign Parts, which sent him Bibles to distribute among the Maroons. Together Chamberlain and Gray worked to convert the Maroons to Christianity, discredit obeah men, and end the practices of common-law polygamy and Coromantee burial rites. They succeeded in baptizing sixty Maroons, a group who formed a separate agricultural settlement at Boydville, about seventeen miles from Preston. Overall, however, Wentworth had to admit that Chamberlain and Gray were ineffectual. The majority of Trelawny Maroons remained antagonistic to all "new ideas," whether religious or occupational.[18]

During these four years Wentworth continually struggled to convince the Maroons that they should stay in the province. The struggle included controlling damage done by Quarrell and Ochterlony, who in February 1797 began to circumvent and undermine the governor's authority in Preston. In April 1797, Quarrell departed for Jamaica. During a stopover in Boston, he received a letter written on behalf of and signed by eight Maroon captains, who beseeched him to convince the Jamaican Assembly to allow them to return to the island. In the meantime, according to Chamberlain, Ochterlony conducted a whispering campaign, reminding the Maroons of the contrasts between "cold barron" Nova Scotia, "fit only for Bears and Moose," and the "paradise" afforded by the warmer climates of Jamaica and Africa. Ochterlony's motivations are murky but he must have been instrumental in writing and sending the Maroons' petitions to Walpole in late 1797. Ochterlony also reassured his wards that Parliament was aware of and responding to their pleas: Walpole and other allies were making preparations to employ the men as patrols in the waters surrounding the Cape of Good Hope, working for the East India Company, or settling in Sierra Leone. Any of these options seemed to the Maroons to be preferable to their current predicament, but Wentworth convinced the captains to try one more year in Nova Scotia.[19] He continued to insist that the Maroons' antipathy to their new home would dissipate as they gradually became habituated to local weather and customs, and especially as they adjusted themselves "to the persevering industry necessary in these Provinces." In 1798 he pointed out that, despite their unabated complaining, after just two years' residence nearly all the Maroons "now weigh two-thirds more than when they arrived here, so fat and lusty are they grown." He replaced Quarrell and Ochterlony with a more trustworthy and stricter superintendent. With the right guidance Wentworth predicted that "the young people will soon fall into the habits of the country and climate" while "the older class will drop away and daily have less exertion and influence." By April 1799, even after more confrontations with the Maroons and more cabbages planted only to be found frozen before harvest, he remained "firmly persuaded . . . that these people are situated in the only part of the world, where they can do no mischief and by a firm temperate perseverance, be reclaimed to civilization and the useful habits of Society."[20]

But the Maroons' own brand of firm temperate perseverance proved stronger—their defiance continually frustrated Wentworth's efforts. In Jamaica, they had sustained a militant culture of organized armed resistance to colonial authority; in Nova Scotia they turned to the arts of passive resistance and petitioning to sabotage the governor's plans for them.[21] On August 6, 1800, the Sierra Leone Company transported 550 Maroons—nearly the entire community, including the Boydville Christians—to West Africa. There they met with Black Loyalists who, though they had previously traveled the same route for many of the same reasons, were now known (without detectable irony) as Nova Scotians.[22] As the directors of Sierra Leone stated in their company history in 1802, both prior to colonization and in a subsequent report to the House of Commons, the Black Loyalists petitioned "to be removed from Nova Scotia on account of the coldness of the climate." Likewise, Wentworth capitulated to the Maroons' petitions and arranged for their departure from Halifax "in conformity to their own repeated requests & solicitations, to avoid the cold winters of this Climate." Whether these groups left the province mainly to escape the cold or whether these complaints were merely an effective means for expressing their grievances against the provincial and metropolitan governments, conflicting perceptions of Nova Scotia's climate were central to both cases, much as they had been in imperial debates about the region's habitability since the early seventeenth century.[23] Wentworth's policy for the Maroons—and resistance to it—must be understood in the broader contexts of British imperial thinking about human and climatic geography in general and northern or cold temperate climates in particular.

Human Geography and Climate

In other parts of the British Atlantic world, those who raised alarms about the dire risks of cold weather for the Trelawny Maroons had never been to Nova Scotia. They took for granted "the supposition that a cold climate is generally understood to be insupportable to negroes."[24] This understanding was informed by long-standing fears of cold northern climates combined with newly relevant, intensifying racial beliefs about their incompatibility with black bodies. But it remained a necessarily general understanding. Through the turn of the century, there was no decisive evidence and no unanimity on the ill effects of inhabiting a cold climate, including its effect on black people's physical fitness.

This uncertainty was partly a reflection of broader transatlantic debates about the correlation between climate and human diversity, debates that both renewed and departed from classical treatises on the subject. Greeks and Romans associated the natural conditions in temperate, tropical, and frigid zones to particular human types, including each type's peculiar susceptibilities to external circumstances, especially daily or seasonal changes in air temperature and humidity. Hippocrates explained in *Airs, Waters, Places*, "you will generally find the complexions and manners of People to correspond with the nature of their Country." Whereas northerners tended to have white or "frosty" features, southerners' skin and hair looked "scorched." The populations of the temperate zone were the most physically diverse, but—as a logical expression of the tripartite climatic order—the majority were supposed to be intermediate in color. Ancient writers further elaborated the causal role of climate in producing geographically specific "humoral constitutions" (determined by the ratio of internal fluids—blood, yellow or black bile, and phlegm). Because Europeans, Asians, and Africans were regularly subject to different degrees of hot, cold, dry, and wet air as well as a variety of wind speeds and directions, these elements caused imbalances—or diseases and "fevers"—common to each locality. Exceptions to such patterns also usually had environmental explanations. Topography, for example, was a variable that could account for anomalies within otherwise climatically and culturally homogenous regions. European "nations that differ in size, shape, and courage" were to be found in "mountainous, rough, dry, and barren countries . . . subject to a great variety of weather." However, if mountain dwellers were subject to the "laws and custom" of their neighbors in flatter lands, they could become less sickly, more courageous, and hardworking. Finally, despite such local variations, humoral imbalances were unavoidable for travelers or migrants to wholly alien environments. Aristotle, Hippocrates, Galen, and numerous other writers' descriptions of climatic geography, human typologies, and humoral makeup (including their lack of system and inconsistencies) proved extraordinary durable.[25]

Borrowing from ideas formulated in Antiquity, early modern writers tended to assume that natural, demographic, and ethnic or physiological variations were mutually contingent. The best proofs for such theories were strong historical patterns of geographical homogeneity, observable throughout the climatic and cultural regions of the Old World. Although ancient writers did not systematically emphasize generic physical traits and their associated maladies over other distinctions or treat them as essential keys to identifying individuals and differentiating human communities, that Europeans were unaware of instances of massive human migrations between the frigid and torrid zones suggested that history could be conflated with the natural order. Until the sixteenth century or later—that is, before European empires had sustained contact with the startling diversity of indigenous populations and ecology of the Americas—the opposite state of affairs was merely counterfactual, largely untested. Conquest, migration, and widespread European and African settlement in the New World simultaneously reinforced and challenged particular aspects of classical biogeography. Early modern writers persisted in idealizing moderate climates and dreading extreme temperatures, especially in equatorial regions where the association between hot climates and African people expanded to include New World slave societies.[26]

In tropical and subtropical colonies, the relative immunity of African slaves to yellow fever or malaria coupled with repeated catastrophic European experiences with these infectious diseases, including the Darien colony in late seventeenth-century Central America, seemed to offer empirical proof of the risks of heat and humidity for the latter and not the former. Eighteenth-century medical attention focused increasingly on understanding and treating medical problems specific to all hot climates, or what came to be known as "tropical diseases." This research on place-specific debility and immunity also contributed to debates among prominent philosophers and naturalists—among them, Linnaeus, Kames, Kant, Montesquieu, and Stanhope-Smith—about whether or not physical differences indicated essential differences in people akin to those they used to classify breeds or species throughout the natural world. Skin color was especially scrutinized. Was color an inherent characteristic or the superficial result of environmental factors like the relative influence of the sun? Virginia-born physician John Mitchell emphasized the latter position when he combined Hippocratic theories of climate and topography, Newtonian optics, and anatomical experiment in his "Essay upon the Causes of the Different Colours of People in Different Climates," published by the Royal Society in 1745.

Well-versed in the physics of light and color, Mitchell reasoned that the "many Degrees of Whiteness and Blackness in the Colours of the People in the World" were a by-product of variations in skin texture and depth. Skin, he argued, is a "transparent membrane" of "different Densities and Thicknesses" that mediates between the inside and outside of a body, variously reflecting or absorbing sunlight. Thick skin appears black, manifesting the absence of color ("Blackness being a negative with regard to Color");

thin skin appears white, showing the combination of the entire spectrum of colors, except when temporarily flushed or sunburnt. Further embellishing his argument with biblical authority and historical conjecture, Mitchell explained that the "original complexion of Mankind" was a "Medium betwixt Black and White" but had diverged over the course of millennia following the Flood. It was possible for changes in climate to produce radical physical changes in people over the very long term but only negligible—if inadvisable—changes in migrants to unfamiliar climates in the short term. As a Virginian, Mitchell could attest that it was best not to tamper with the climatic geography of race: "White People are most healthy in cold and black or tawny People in hot Countries; each being subject to Disorders, on a Removal to these respective Climes." In addition, Mitchell argued that quick changes in color would always be more apparent in white-skinned people: Dark-skinned people did not lighten when they moved north for the same reason that it was easier to saturate white cloth with color than the reverse. Amplifying some aspects of Mitchell's theorizing, Jamaican planter Edward Long pointed to the black populations of New England and New York, "where the winters are more severe than in Europe," to demonstrate that migration to foreign climates never changed skin color or any of the other physical, mental, or behavioral traits of "Negroes" that he listed in scabrous detail in his *History of Jamaica* (1774).

Such strident assertions were representative of an increasingly forceful ideological position—but not a consensus or orthodoxy—on theories of racial fitness. Even as the Atlantic slave trade and slavery structured social relations and justified perceptions of a universal racial hierarchy with fair-skinned, fair-minded "Caucasians" at the pinnacle, the medical geography of race was still subject to widespread speculation. As historian of medicine and empire Mark Harrison argues, through the early nineteenth century "most writers on fever upheld received wisdom on seasoning and acclimatization" rather than "innate differences" for explaining relative vulnerability to the health hazards peculiar to different climates.[27]

Physicians and administrators devoted much less attention to understanding the specific disease environments of cold-climate colonies, even after the Seven Years' War when the empire came to encompass a vast expanse of territory in North America's northern latitudes. This neglect reflected the knowledge that infectious diseases affecting Indians and colonists in this region were the same as those common to Britain and Europe—measles, various forms of pox, scarlet fever, scurvy, and so on. It implied that a relatively familiar environment was a relatively salubrious one for travelers. As a British traveler wrote about Nova Scotia in 1765, "its northerly situation exposes it to severe cold and deep snows in winter; but is generally very healthy, and agreeable to English constitutions, as are all the northern provinces." Scottish physician to the East India Company James Lind's influential *Essay on Diseases Incidental to Europeans in Hot Climates* (1768) was mainly about the subject announced by its title, but he began with advice about "diseases incidental to strangers in different parts of the world." These remarks included passages on the climates of a somewhat miscellaneous list of British American colonies, including Quebec, Nova Scotia, and New England (taken as a whole)—all places he judged as healthy as the home country, especially compared with Maryland, Virginia, and South Carolina. About Nova Scotia, he wrote that "the long-continued health enjoyed by those who pass the winter at Halifax, are proofs that an intense degree of cold, properly guarded against, produces but few diseases and scarcely ever the fevers which are the subject of this treatise"—an endorsement strikingly similar to Wentworth's remarks three decades later.[28]

Although Nova Scotia's cold climate produced no exotic diseases, skeptics believed it caused stagnation and decline. The famous tract *American Husbandry* ridiculed the idea of any viable life in the province. During its seven-month-long winters, "the inhabitants are shut up in their houses, and . . . lead a miserable life; are almost in as torpid and lifeless state as the vegetables of the country. Such a degree of cold as is then felt benumbs the very faculties of the mind, and is nearly destructive of all industry." Similar condemnations of Nova Scotia (and its relative value to the empire) proliferated after each Atlantic war that reaffirmed British possession of it. In 1713, 1763, and 1783, fresh reports provided a more detailed and complicated picture of the colony's environment, but usually to the further detriment of its reputation. By the late eighteenth century, even northernness came to matter less than locality. Quarrell's suggestion to send the Maroons to Upper Canada, for example, was justified not only for confining them to a landlocked situation but also because it was believed that the interior of the continent was surprisingly warm or, "at least the climate . . . was sufficiently temperate for the plan proposed." The fact that Upper Canada was also in northern latitudes was misleading because its climate was "by no means to be compared for severity with that of . . . Halifax."[29]

Responding to such statements, Titus Smith, a botanist and land surveyor in Nova Scotia, argued that, "unlike Lapland or Sierra Leone," no one would "degenerate by the fault of the climate" in Nova Scotia. In choosing these examples, Smith acknowledged the standard framework of medical geography in which Lapland and Sierra Leone figured as places in the frigid and torrid zones where nonnatives became sick from relentlessly severe weather or died from exposure. By denying the equation with Lapland, he implicitly diminished the climatic differences between Nova Scotia and other temperate regions—a typical strategy of rhetorical climate change. Titus Smith, who with his parents had moved from New Haven to Halifax at the start of the American Revolution, maintained that Nova Scotia's winters were not only tolerable but among its natural advantages. Settlers satisfied with life in the region construed its thin soils and cooler climate as positive features. Although "there are certainly countries from which a prudent man would emigrate if possible, the more sterile part of Nova Scotia is not one of them." Reversing Buffon's theory of New World degeneration, Smith went further: He proposed that acclimation to Nova Scotia's cooler winters and thin soils promoted health. "That this country is in some degree indebted to the barrens for the salubrity of its atmosphere there can be no doubt." Challenging environments were invigorating, whereas hospitable climates and bountiful landscapes were demoralizing:

> Wherever there is an unusually large district of very fertile land the inhabitants are unhealthy and liable to degenerate and become a weak timid race, who require a constant supply of men from other countries to support their population. While the inhabitants of a poorer soil who are compelled to use greater exertion to procure a subsistence have more strength and energy.

Nova Scotia's brisk winters, he concluded, offered "a very healthy climate to the temperate" and a bracing environment for the "intemperate."[30]

At the same time, in choosing these examples Smith was likely responding not only to the polemics inspired by Buffon's assertions about cold climates but also more specifically to unfavorable comparisons between the climates of Nova Scotia and Sierra Leone that were generated during the 1780s and through the turn of the century on behalf of the Sierra Leone Company. In his 1786 proposal for a British settlement colony in Sierra Leone, the naturalist Henry Smeathman argued that this part of the West African coast would be particularly suitable for the empire's poor "Blacks and People of Colour," who would form the new colony's main constituency. Unlike the climate of Nova Scotia, the "mildness" of Sierra Leone would be "congenial to their constitutions." According to Smeathman, the difference between the two colonial environments was absolute: Sierra Leone's "most pleasant, fertile climate" was "fit and proper," whereas Nova Scotia's was the opposite: "unfit and improper."[31] Smeathman derived his conclusions on the climatic frontiers of racial health from comparative fieldwork on what he called the "intertropical zone" of West Africa and the Caribbean, within which plants, insects, and animals, including people, could supposedly be transported without need for acclimatization (a later analog of Vetch's notion about migration between cold environments). It was a version of environmental determinism, which suggested that black bodies required and had to be limited to warm climates. The idea was implicit in Jeremy Belknap's response to Virginian St. George Tucker's queries about the health of free blacks in Massachusetts after their emancipation in 1790. Belknap thought that African Americans had never been numerous in Massachusetts during the colonial period and those who did live there were "always more sickly and died in greater proportion. The winter here was always unfavourable to the African constitution."[32] In his natural history of the Arctic, Thomas Pennant made this idea explicit in a converse example, when he explained that the Inuit in Greenland were "a race made for the climate, and could no more bear removal to a temperate clime, than an animal of the torrid zone could into our unequal sky: seasons and defect of habitual food would soon bring on their destruction."[33] So too, Walpole had argued that the Maroons would always be at a natural disadvantage in Nova Scotia.

Climatic determinism could thus offer a convenient, compelling, and apolitical explanation for the poverty and poor health typical among blacks in northern latitudes, from southern New England to London. At the same time, racial climatic determinism was an exceedingly inconvenient idea for governing settler populations in a global empire, particularly in tropical regions. If the Sierra Leone Company assumed that their West African settlement's working majority would be black, they also believed that its ruling minority should be white Britons, some of whom would have to reside in the colony. Smeathman admitted that the West African climate had "been fatal to many white people," but he defensively argued that these fatalities were due to recklessness: pitching camp near swamps, leading "most intemperate lives," eating unhealthy food, and drinking too much liquor. For disciplined white settlers, the climate would prove "very healthy," so long as they limited their alcohol consumption to moderate amounts and settled arable lands where they could grow and eat local foods.[34] In 1802, Zachary Macaulay, secretary and former governor of the Sierra Leone Company, updated Smeathman's medical advice for "Europeans," who must have continued to worry about relocation to West Africa. On first arrival even the Nova Scotians and Maroons were discomfited by Sierra Leone's heat and humidity, but they soon adapted. As one of Macaulay's officers noted, "remembrance of the cold of Nova Scotia" was "a sufficient check" to the Maroons' complaints. White newcomers would also adjust to the heat and "enjoy a good state of health" as long as they were "not exposed to excess of labour" and took extra precautions during the rainy season.[35]

Thus there was no definitive answer to the question of whether Nova Scotia's climate would be any "less tolerable" to the Maroons "than other Blacks." Even avowed racists like Jamaican planters Bryan Edwards and Robert Sewell judged the idea of racial climatic determinism to be too reductive. Walpole's motions in the House of Commons on behalf of removing the Maroons from Nova Scotia to a warmer climate were ineffective partly because Edwards, Sewell, and other opponents pointed out that the Trelawny Maroons had mostly occupied Jamaica's mountainous highlands, where temperatures were 20° to 25° cooler than in the lowlands, "nearly as cool as the mountains of Liguanga." Going still further, Sewell's ally Joseph Barham provocatively likened the Maroons' experience of a "change of climate" to "no more than what was experienced by many gentlemen, either in the performance of their duty, in attending to their affairs, or in gratifying their curiosity." This comparison suggests a surprising leveling: With enough determination, the Maroons might do as well in the cold North as genteel white Europeans did in equatorial regions—even if both classical tradition and recent history suggested that neither group fared particularly well in extreme temperatures.[36]

While climatic geography was increasingly connected to emerging scientific ideas about race, questions about the nature, origins, and fixity of human differences remained unresolved. In the colonies, officials rehearsed the clichés of racial climatic determinism when it was opportune and ignored or modified them when it was not. During the Revolutionary War, Lord Dunmore suggested that arming slaves in South Carolina made sense in part because they were "better fitted for service in this warm climate than white men"; after the war, black soldiers were sent to live in the colder climate of British North America irrespective of the fit.[37] In settler colonies like Nova Scotia and Sierra Leone, convictions about the climatic geography of race were usually loosened by the more abiding problems of encouraging and maintaining a sufficiently large, permanent population of farmers and laborers.

Wentworth on Climate, Slavery, and Race

Wentworth's solicitude toward the Maroons does not mean that he wholly rejected climatic theories of race. At no time did he dispute that blacks would suffer in extremely cold, northern environments. His contention was that there was no extreme cold in Nova Scotia, an argument that was largely informed by his comparative sense of New Hampshire's climate. His family had dominated New Hampshire's colonial government from the early eighteenth century through the Revolution. Following his grandfather John and his uncle Benning, Wentworth served as royal governor of New Hampshire and surveyor general of the King's Woods from 1766 to 1775. From this "long experience in New England" he believed that the winters were "more severe than in Nova Scotia" and yet these conditions had been harmless to the region's free and enslaved black residents. "The idea that our own climate will distress them, maybe started: But, believe me, Sir," he wrote about Nova Scotia in 1796 to the British minister in the United States, "long & extensive experience completely in point in N. England & in this country, justifies me in assuring you that the apprehension is specious only." He was "satisfied there need not be any apprehension entertained of this Climate injuring them." Instead, he suggested that exposure to chilly temperatures was beneficial for newcomers, particularly for blacks. In New Hampshire he had "always found Negroes directly from the hottest coasts of Africa have grown strong and lusty in the Winter."[38] Wentworth quipped that acclimation to the North's milder weather would have a particularly salutary effect on the Jamaican Maroons. In combination with Gray's and Chamberlain's teachings, cool weather would serve to "eradicate all former savage ideas & substitute more mild and happy influences." The moderate temperatures of Nova Scotia would season the Maroons in spirit as well as in body, cooling their fiery temperaments and tempering their violent inclinations.[39] Conversely, when it became clear in late 1799 that the majority of Maroons would soon depart for Sierra Leone, Wentworth cautioned Portland that African heat would inflame their natural volatility: "All the men and youth will require great circumspection, vigilance and power, to govern them in that climate, where they will be no better disposed to earn their own subsistence than in Nova Scotia." He presumed that British elites agreed on such climatic influences. That is why, given the options, he had been certain that his provisions for the Maroons in Nova Scotia's temperate climate would "satisfy Mr. Wilberforce among other reasonable philanthropic Patrons of the black race."[40]

Wentworth never explicitly revealed whether he considered himself to be one such patron, an ally to Wilberforce's abolitionist cause. From the seventeenth century until slavery was abolished throughout the British Empire in 1833, the northern colonies were societies with slaves. During the periods when Wentworth was governor of New Hampshire and of Nova Scotia, slavery was legal or quasi-legal in each province. For economic rather than political or moral reasons, New Hampshire and Nova Scotia happened to be the only British colonies in which slaveholding did not expand before the American Revolution. After the war, however, slaveholding and indentured servitude increased dramatically throughout the Maritimes with the influx of thousands of white refugees who relocated with between 1,200 and 2,000 African American slaves. It must have seemed a bitter irony to the emancipated slaves who were veterans of war that this retreat provided, at best, an imperfect sanctuary.[41] The continuation of slaveholding in Nova Scotia through the turn of the century reinforced the common equation between African physical features and slave status, which in turn exposed free blacks to racial discrimination and threatened them with the prospect of reenslavement. This threat was heightened by the legal ambiguities surrounding slaveholding in the province. There were no statutory laws explicitly sanctioning or abolishing slavery, slave sales were rarely recorded, and white loyalists often referred in documents to their domestic slaves as servants. And yet, white settlers could purchase slaves in Halifax and slaveholding remained widespread. Particularly in the unstable and complex period immediately following the Revolutionary War, racial attitudes tended to trump legal status in Nova Scotia. Notwithstanding irregularities in provincial law, white settlers regarded slavery as a customary privilege, a privilege they could enforce by resort to local property laws that implicitly regulated the ownership of human chattel. Inconsistencies between statutory and common law were exposed in petitions raised by black settlers, abolitionists, and slaveholders; court cases related to the legality of reenslavement and disputes over slaveholders' rights to or compensation for losses of human property; as well as a move by a provincial assemblyman in 1801 to establish a public inquiry into "the rights which Individuals in the Province have to the Service of Negroes and People of Colour as Slaves; and also to ascertain the Value of such Slaves, that a Sum of Money be appropriated to pay such Individuals for their Property in such Slaves."[42] As one freeman stated in a petition to the provincial assembly in 1790, free or indentured "people of colour [were] injured . . . by a public and avowed Toleration of Slavery."[43]

Racial discrimination between black and white refugees manifested most sharply in differential access to farmland. Only a fraction of the 3,000 Black Loyalists sent to freedom in Nova Scotia were granted land, and those who received lots found them to be smaller than those granted to whites. Grants to Black Loyalists ranged from between one-eighth of an acre to 750 acres compared with a range of between hundreds to tens of thousands of acres granted to white loyalists. Moreover, the Black Loyalists' grants were concentrated in Preston and Birchtown, new towns established in areas along the colder Atlantic coast with some of the most sterile soils in the province. James Hamilton, one of the more substantial white landowners along the Cape Negro River to the southwest of Birchtown, told surveyor Titus Smith that "we should not find a foot of land between his House & Argyle, that the greatest part of the Land towards Barrington & Argyle was covered with Rocks & the Remainder like the Land near his House which is a poor Land covered in general with Pine."[44] In 1785, William Booth described the Black Loyalist settlements in Birchtown as a "valley with much stones and a little swampy." These factors made even subsistence farming untenable on most of the lands granted to Black Loyalists and pushed many of them into lifelong debt peonage to whites in better standing. Reflecting on the situation of the Black Loyalists in Nova Scotia, Joseph DesBarres surmised that they could only have agreed to settle in the province as a desperate and temporary resort.[45]

For individuals without property of any kind, maintaining liberty or self-sufficiency was even more difficult. Reenslavement for free blacks was a common danger. The misfortunes of Lydia Jackson, a Black Loyalist from South Carolina, illustrate the precariousness of black freedom in Nova Scotia. The abolitionist John Clarkson, who was in the province in 1791 on behalf of the Sierra Leone Company, recorded her story. Although Jackson was a free woman, she was destitute, pregnant, and had been abandoned by her husband. A white loyalist deceived her into signing a thirty-nine-year contract of indenture and sold her (under the terms of this contract) for £20 to a Hessian veteran and his wife in the German settlement of Lunenburg. Essentially a slave, her masters regularly and severely abused her, even while she was pregnant. But Clarkson told her that, because her problems were widespread among Black Loyalists throughout the region, she had little hope of legal redress. The following year Jackson left with Clarkson for Sierra Leone.[46]

In considering whether the Maroons would face racial hostility comparable to that of the Black Loyalists' experiences, Wentworth acknowledged that, although slavery was "almost exterminated" in Nova Scotia, "distinctions naturally painful to these people" were only "gradually dying away."[47] But he argued that it was primarily because of suffering from bigotry and economic disadvantages—rather than dislike for, unfamiliarity with, or unique vulnerability to the weather—that prompted almost 1,200 Black Loyalists (over one third of the original number of refugees) to leave for Sierra Leone in

1792. Some individuals, like a man who said he would rather die in "his own country than this cold Place," pointed to the winter as a reason for emigration to Africa. But Wentworth resented that dishonest men like Ochterlony had "imputed" most individuals' decisions to leave the region "to the only cause which had been really friendly to them—viz.—The Climate." That the climate was not the main issue, he argued, was proven by the fact that the majority remained. Observers outside of Nova Scotia corroborated the point by noting that those "families of negroes" who stayed in the province were "enjoying, as farmers, comforts equal to those of their white neighbours." This was a misleadingly rosy view of most people's circumstances or autonomy after 1792—besides slaves and indentured servants in the province, Wentworth knew that most black freeholders continued to be too poor to properly care for their livestock or even to clothe themselves. Most relied on annual subsidies from the provincial government in exchange for men's militia service as infantry laborers. Despite these hardships, Wentworth maintained that he had "never in any instance heard them complain about the Climate."[48]

Other than such statements, which he made directly in response to the transatlantic debate about the presence of Black Loyalists and Jamaican Maroons in his province, it is unclear exactly what Wentworth thought about questions of race or slavery, the slave trade, and their abolition. Despite Wentworth's multiple honorary degrees in law and letters and his initiative in founding Dartmouth College, he and Frances were known more for their lavish parties rather than for sustained political activism or intellectual engagement. Like other gentlemen of his rank, Wentworth collected books and was curious about natural history, but when he left New Hampshire the citizens auctioned off his pre-Revolutionary library, and there is no record of what was in his collection at Nova Scotia's Government House. He must have been aware of the antislavery pamphlet *Letter to a Clergyman Urging Him to Set Free a Black Girl He Held in Slavery*, written and published in Halifax in 1788 by Presbyterian minister and Highland emigrant to Nova Scotia, James MacGregor. In addition, during the 1790s Wentworth revived his friendships with fellow Harvard graduates Jeremy Belknap and John Adams, both of whom were public advocates of gradual emancipation and the abolition of the slave trade in the United States. But none of their extant correspondence with Wentworth touches on these subjects. As a result, it is difficult to know when (or even if) Wentworth's personal views and policies in regard to antislavery shifted from indifference to what could be characterized as his passive sympathy for northern blacks.[49]

Wentworth's record of personal dealings reinforces the indeterminacy of his position in regard to slavery or ideas about racial difference. There are no pre-Revolutionary records indicating that Wentworth owned slaves, as did many members of his extended family in New Hampshire, the West Indies, and Britain. In 1783, however, Wentworth purchased two musicians named Romulus and Remus in New York as a gift for Lady Rockingham.[50] The following year, he brokered the purchase and shipment of nineteen slaves (whom he praised as "a most useful lot of Negroes") from Halifax to his cousin Paul Wentworth's estate in Surinam.[51] Another window onto Wentworth's racial attitudes was his intimate relationship with the Maroon woman Sarah Colley, who worked as a domestic servant at his summer estate in Preston and bore their son, George Wentworth Colley. Little is known about Sarah Colley except that, given that George was born in 1804, she did not leave with the rest of the community in 1800, and George's descendants continued to live at the Preston estate into the twentieth century. There are many contemporary examples, both infamous and obscure, of similar relationships between North American officials and black or mixed-race women in the eighteenth century, from Thomas Jefferson and Sally Hemings to Walter Patterson, the first governor of Prince Edward Island, and his "mulatto mistress" Susanna Torriano, who inherited from him substantial landholdings on the island. If Wentworth's affair with Sarah Colley was not unusual, it does suggest a still more complex facet of his objections to the removal of the Maroons from Nova Scotia.[52]

It is also possible that his equivocal views of the Maroons reflected their exceptional political and social history in relation to other black people in the British Empire. The outcome of the Second Maroon War showed their ultimate vulnerability to the colonial government, but prior to the Trelawny Town Maroons' deportation from Jamaica they represented a living paradox of racial slavery—privileged but vulnerable "islands of freedom in a sea of slavery," in historian Richard Sheridan's words. Like maroon communities throughout the Americas, Jamaica's five groups were formed by fugitive slaves who had succeeded in resisting reenslavement, a status that was formalized in the Treaty of 1739. At the same time, the treaty stipulated that their independence was contingent in part on helping to reproduce the plantation regime from which they had escaped. Maroons developed a complex military hierarchy but depended on the colonial government for ammunition; they periodically raided plantations or provoked slave rebellions while generally acting as slave catchers who reinforced masters' domination over slaves.[53] Jamaican Maroons came to define themselves as politically, ethnically, and culturally distinct from slaves. That Jamaican elites accepted this distinction was reflected in Edward Long's proposal in the 1770s to conduct an annual census of the island on which Maroons would be counted in the same category as soldiers, free blacks, and mulattoes (other categories included Negroes and other slaves, livestock, absentee landowners, and white settlers of all ages and genders).[54]

For most local elites in Nova Scotia, however, the Maroons' exceptional history was largely irrelevant or at least obscure. When the provincial assembly formed a special committee in 1797 to ensure the governor kept them apprised of the source and amount of funds spent on behalf of these fresh arrivals, they referred to them as "a certain Description of Black Persons, known by the Appellation of Maroons." The assemblymen were adamant "to prevent such Black Persons from eventually becoming chargeable to the County of Halifax, the Place of their present Residence."[55] Wentworth also sometimes made this equivalence between the Maroons and other people of color residing in the province. Like many abolitionists whose humanitarian activism was guided by deepening certainty about the significance of physical differences, there was little reason for him to rigorously distinguish the Maroons from Nova Scotia's free blacks, slaves, or Indians. Maroons could be lumped with other darker-skinned people, all lesser brethren in need of moral reform and protection, which might include protection from the supposed ravages of alien environments. In this sense, the ambiguities and inconsistencies of Wentworth's thinking about the Maroons in relation to other nonwhites were typical of late eighteenth-century attitudes. In hindsight, the inconsistency of racial thinking seems particularly glaring at the turn of the century, a period during which, as historians have shown, the meanings, implications, and application of racial ideology—including belief in biological racial difference and paternalistic or missionary concern for nonwhites—were only beginning to emerge in forms that are readily recognizable from a modern vantage. Just as arguments for polygenesis and the enduring features of human physical difference were gaining in influence, so too were sympathetic arguments for emancipation or, less radically, measures for ameliorating the health of slaves.[56] For officials throughout the empire, despite the growing authority of science, the utility and deployment of racial theory remained highly contingent on immediate circumstances. Colonial officials like Wentworth set aside or softened any rigid adherence to definitions of racial difference if some other form of reasoning proved more expedient when pressing concerns impinged on relations between whites and blacks or between administrators and subjects.

Empire Without Slaves in the North

Although Wentworth's actions showed some aspects of antislavery impulses, they are mainly comprehensible as policy responses to colonial problems, policies that must be contextualized in the broader history of immigration schemes in this period. Above all, it is crucial to understand how racial thinking about climatic geography coalesced with some of the earliest expressions of abolitionist reforms embedded in immigration schemes. Most of these schemes were proposals hatched after the Seven Years' War for populating new but undeveloped imperial territory in hot climates from the Gulf of Mexico to Africa. In these territories, abolitionism and climatic determinism formed the justifications for a number of colonial settlement projects for free blacks—from an idea floated in 1763 to offer freeholds to Christianized Africans in Pensacola, Florida, to the foundation of the Sierra Leone Company, which justified the transportation of over 1,600 free blacks from London and Nova Scotia to West Africa in 1792, in part because blacks were supposedly "accustomed to labour in hot climates."[57]

Although central figures of British abolitionism like Wilberforce and Granville Sharpe were principals in the Sierra Leone Company, most such schemes were not promulgating a radical antislavery agenda. Instead, immigration and settlement schemes with abolitionist elements were typically promoted by members of a broad, loose coalition of interests in favor of free labor both in Britain's industrial factories and in its settler colonies. This is because, before Britain's abolitionist movement turned from a shared moral sentiment that slavery was evil to concerted action against it, abolitionism initially arose out of an unlikely and incoherent alliance of people with disparate interests. During the last decades of the eighteenth century—a particularly volatile and experimental period in imperial governance—some individuals became involved in the abolitionist movement because it fit with their broader political or financial ambitions in the empire whereas others joined the cause because of highly parochial interests related only indirectly or not at all to antislavery or concerns about the welfare of slaves or black people.[58] George Walpole, for instance, was entirely willing to use bloodhounds to defeat the Maroons in Trelawny Town; his concern for them arose when the Maroons' bad fortune intersected with and jeopardized his own reputation. The abolition of slavery and the slave trade was thus not just a lofty ideal but also practical politics and economics. As Olaudah Equiano explained in his *Interesting Narrative* (1789), the cause should appeal not only to slavery's victims and their humanitarian advocates, but also to British industrialists and merchants for whom it would open Africa as "an endless field of commerce." He encouraged Smeathman's initiative for free black agricultural settlements, where tropical plantation crops like cotton and indigo would be grown and exported by British Africans, who would in turn become consumers of finished goods from Britain. Equiano reasoned that, because the British Empire's "manufacturing interest and the general interests are synonymous," it followed that "the abolition of slavery would be in reality an universal good."[59] A compelling—because paradoxical—proof of this broader view of abolitionism's political currency is that it also won approval among some quarters of London's proslavery lobby. Absentee West India planters were among those who heartily supported not only amelioration but also the inclusion of slaves as part of Britain's imperial community of subjects, with some of the attendant rights. Others were champions of the city's Committee for Relieving of the Black Poor. To demonstrate that their benevolence toward impoverished urban blacks was compatible with their interests in the continuation of the slave trade and Caribbean plantation slavery, some absentees were in favor of creating free black settler colonies on the slave-trading coast of West Africa.[60]

The first concrete, if little-known, scheme for gradual emancipation was described in the obscure pamphlet *Plan for the Abolition of Slavery in the West Indies*, which circulated in manuscript form immediately after the Seven Years' War and was published in 1772. Maurice Morgann—who served as secretary to the president of the Board of Trade and in colonial posts in Quebec, New Jersey, and New York—was the *Plan*'s likely author. Morgann's *Plan* proposed removing West African children to England until the age of sixteen, thereafter transporting them to Pensacola in British West Florida to develop a free black agricultural settlement. If Morgann's proposal represents an early formulation of abolitionist ideas, it was also one that was thoroughly committed to climatic theories of race. To "whoever attends to the economy of nature," he wrote, it should be self-evident that Florida would thrive best if planted by blacks because "the natives of Africa are better able to bear extreme heat, and to labour in the Torrid Zones than the whites of Europe; and that the former may, under the influence of a right policy, be induced to emulate, in one climate, the industry of the latter, in another." Morgann then described the future landscape of gradual emancipation in Britain's settler colonies. While black workers would populate the Pensacola settlement and other southern regions of British America, he envisioned the northern colonies populated by low-investment white workers: "The whites will inhabit the northern colonies and to the south the complexions will blacken by regular gradation." Only in "the middle parts," would whites and blacks intermarry and thereby "link the two extremes in union and friendship." Although he did not specify the exact locations or extent of "the middle parts" (a vagueness typical of references to the geography of the temperate zone), he did consider the possibility that his emancipation plan could also be pursued, if "imperfectly," by transporting children from Africa or the West Indies "into the northern provinces of America, to be there educated and from thence transplanted to Pensacola." In a more ominous passage, however, in which Morgann predicted an imminent "war between the races," he warned that settling blacks permanently in the northern colonies would be a grave mistake. There would be "no truce, no cessation of rage, till the violated order of Nature is restored, and the climates of that country are divided between their proper and destined inhabitants." To respect the economy of nature, including the fact that racial differences were "real and lasting," the empire needed to create a geographical division of labor and settlement. Morgann offered a comprehensive vision of what historian Richard Drayton has aptly called a "nature's government"—in this instance, an imperial order rationalized in terms of climatic geographies of race that would ultimately result in a more just, stable, and efficient empire.[61]

Although Morgann's proposal was the earliest to specifically and explicitly promote abolitionism, it was preceded and influenced by other schemes for marrying the aims of empire with philanthropic overtures, treating colonial planning as a project of moral uplift. The most prominent eighteenth-century example was the original plan for the colony of Georgia, which provided transportation, asylum, and small land grants to the English poor, debtors, and orphans. In stark contrast to the plantation economy and black slave majority of its neighbor South Carolina, large landholding, slavery, and rum were banned. According to the scheme, temperateness would inhere in Georgia's settlers and the agrarian landscapes where they were expected to cultivate Mediterranean products like silk and wine. Such presumptuous charitable endeavors also underpinned projects for producing permanent settlers in the North. In 1749 (coincidentally the year that Halifax was founded), the Massachusetts physician and pamphleteer William Douglass suggested that the northern colonies should follow the example of the Society for the Propagation of the Gospel's missionary schools in Ireland, which leased farmland "for the Improvement of the Boys in Husbandry, and for the Profit of the School." Douglass thought Americans could reform Indians and "poor Negros" using this same approach, removing youths from their families or villages and putting them to work on small farms adjacent to a school.[62]

Whether Wentworth was aware of Morgann's Pensacola plan or Douglass's proposal, he worked within the basic outlines of such schemes in crafting a similarly creative solution for managing and expanding settler colonialism in his underdeveloped corner of the British Empire. An empire based on, or at least encouraging of, disciplined agricultural laborers aligned neatly with the interests of the northern colonies and seemed to offer multiple mutual benefits. The peninsulas of Nova Scotia and Florida were both regions of British North America that had been underfinanced and underpopulated, with insufficient funds to significantly invest in slavery. They faced chronic labor shortages and difficulties in encouraging permanent settlement.

Although Wentworth had anticipated the problem of attracting permanent settlers, he had underestimated the difficulties he would face in solving it. On his first arrival to the province in 1784—the moment when Halifax and other towns were overwhelmed with Loyalists living in temporary housing—he predicted that, until the white loyalists "are on lands actually their own they are only transient, have nothing to bind & cement them, and many will again migrate." The surest way to retain them in the province would be to distribute land "as soon as possible" and "to get them employed in Agriculture." He was confident that Nova Scotia would be "peopled in a very short time" with improving settlers, "those whose habits, occupations, fears, sufferings, prudence, spirit, zeal, abilities, and industry, would make this [colony] a secure & inestimable Treasure to Britain." A decade later, however, economic growth in the province was due mostly to wartime subsidies rather than to stable population increase or immigration.[63] The province could economize its disbursement of these subsidies by exploiting the surplus labor of Maroon and Mi'kmaw adult men for military service. In the meantime, Chamberlain and Gray would coax their children "as they grow up" into accepting Christianity and the farming "lifestyle." Once reeducated, Wentworth believed that the Maroons would "honestly perform" their duties to the colony as well as could be expected from "more enlightened white people, from any part of Europe or America, far more easily reformed."[64]

Incorporating indigent, ethnically marked populations like the Black Loyalists, Trelawny Maroons, and Mi'kmaq could seem outlandish, but it could also be a timely solution to a range of fundamental colonial and imperial problems. Offering independence and land grants to these groups would reduce the likelihood that they would attack white settlers or revolt against the colonial government. Teaching them the rudiments of good husbandry would also ease their dependence on the public purse and advance the cause of agricultural improvement. However, most local elites only partly agreed with this vision, particularly because they saw the Maroons merely as a cheap source of labor, with as little claim to special treatment or rights as other constituents of the provincial underclass. White landowners believed that assimilating the Maroons to local customs was crucial, but did not want to compete with them for arable land and were more sympathetic to Quarrell's argument that "dispersing the Maroons very extensively, was the only means of disposing of them properly . . . even to spread and extend them in small settlements as far as New Brunswick."[65] Going further, some provincial elites insisted that the Maroon community must be entirely dismantled. Rather than allotting the Maroons their own tracts (whether dispersed across the region or in a clustered settlement near Halifax), families should have been separated and individual men, women, and children put to work as indentured laborers on the large estates scattered along the northern and western coasts of the peninsula, where they would be employed as field hands or domestic servants. Simultaneously mollifying the Maroons, assuaging metropolitan abolitionists' vague humanitarian concerns for them, and appeasing local elites made Wentworth's scheme impracticable. Ultimately, he was unable to negotiate a compromise between the demands of these irreconcilable local and transatlantic interests and his policy experiment was short-lived.

Its failure was due in large part to the complexities of ideas about race and climatic determinism. In the late eighteenth century, even as these became increasingly compelling concepts for understanding or justifying the differential treatment of people, they never uniformly guided policy-making across the empire. This is because climatic geographies of race did not simplify long-standing administrative problems of attracting settlers to, or managing populations within, individual colonies. Indeed, in both Nova Scotia and Sierra Leone the emergence of biological racism tended to further complicate or exacerbate rather than to solve such problems. The Jamaican Maroons' forced exile in North America and Africa were the outcome of several competing factors, among them turbulent social relations in plantation colonies, the constraints on settler colonialism in what were perceived to be extreme environments, and the vagaries of racial thought. That Wentworth's policy experiment failed to harmonize these variables put him in the company of numerous other colonial officials, from James Oglethorpe of the Georgia Trustees to the governors of Sierra Leone.

Nevertheless, settling the Maroons in the northern climate of Nova Scotia provided a kind of test case for theories about the relationship among climate, geography, and race and its intersection with the political economy of empire. For committed determinists, defying the climatic geography of human difference was plainly wrong. Others were persuadable but more tentative. For example, some perceived that black people were hardier in all conditions. As one member of Parliament put it, their "nature seems better able to bear the severity of cold, than the whites can that of heat." But neither combination was advisable. Notwithstanding his "high opinion" of Wentworth, nothing could make "people happy who were taken from the burning heats of Jamaica, and exposed to the severe colds of Nova Scotia."[66] Conversely, they could also point to a long history of ill-fated attempts to transplant whites to southern climates, including the Darien misadventure; the French Empire's failed Kourou colony in northern South America, which was briefly settled by Acadian farmers after their deportation from Nova Scotia in 1755; or the still more tragic scheme to transfer Greek peasants to British East Florida, where most died within months after their arrival in the 1760s. These three particularly egregious examples were among the many cases that seemed to prove that tropical and even subtropical environments were deadly for whites.[67]

For local elites in Nova Scotia, the lesson of these debacles was more subtle. Many officials in British North America were sympathetic to the causes of abolishing the slave trade and slavery or of gradual emancipation, in part because they embraced antislavery activists' climatic explanations about the differences between whites and blacks. More important, the transatlantic and parliamentary debates about Maroon settlement in Britain's northern colonies did not resolve uncertainty about whether or not the region was in the temperate zone. For those invested in peopling the province, the heightened attention to the region's climate in the late 1790s was an opportunity to prove that, crudely stated, if black people could live there, surely any sort of white settler could too.

Not everyone who supported the removal of the Maroons from Nova Scotia fully disagreed with them. The jurist Robert Charles Dallas, an abolitionist who wrote the first history of the Maroons' Atlantic odyssey immediately after they arrived in Sierra Leone, argued that their suffering from the climate in Nova Scotia had been a matter of degree. In principle, he believed that whites were more adaptable than blacks to unfamiliar circumstances. But he also knew that this difference could be indiscernible in practice: In the "very sharp" winters of New York, New Jersey, and Pennsylvania, for example, "negroes are mixed with the white people as labourers." Because blacks lived throughout the colder temperate climates in North America and Europe, he concluded that they could probably "exist and thrive in every region inhabited by white people." But, he added, "whether it be equally pleasing is another question."[68]

5

Works in Progress

Both the heat and cold are now far more moderate, and the constitution of the air in all respects far better, than our people found it at the first settlement. The clearing away of the woods, and the opening of the ground every where, have, by giving a free passage to the air, carried off those noxious vapours, which were so prejudicial to the health of the first inhabitants. The temper of the sky is generally, both in summer and winter, very steady and serene. Two months frequently pass without the appearance of a cloud; and their rains, though they are heavy, do not continue any long time.

ENTRY FOR NEW ENGLAND in "The Philosophical and Political History of the Thirteen United States of America" (1784)

WELL BEFORE THE fiasco of the Maroon deportation, Joseph DesBarres declared that "the climate is perhaps the greatest natural Evil attending this country." As soon as "the interior of the country comes to be settled," however, he had "no doubt but this evil will diminish here as it has elsewhere." Like Montesquieu, David Hume, Benjamin Franklin, and Thomas Jefferson, DesBarres subscribed to a neoclassical theory of anthropogenic climate change. Combining ancient history with contemporary observations and geological science, numerous learned elites insisted that Europeans and colonists had for better or worse changed the climate in their localities. If Greco-Roman Empires had maintained or improved the Mediterranean climate, their decline initiated a centuries-long process of desertification. These changes seemed to be aloft and advancing much more rapidly in territorial peripheries throughout the early modern world, where different forms of land management resulted in the deterioration or amelioration of local climates. In tropical plantations, naturalists were troubled by widespread deforestation, which left the ground overexposed to direct sunlight and winds and intensified the desiccation and erosion of topsoil resulting from monoculture.

On the other hand, in settler colonies, farmers' work in displacing or assimilating indigenous populations, converting forests and swamps into arable land, and expanding the frontiers of commercial, civil society had perceptibly and measurably softened harsh climates—at least according to some early naturalists' calculations. From the Scottish Highlands to Vermont, cold climates warmed. From West Africa to South Carolina hot climates cooled. DesBarres believed that the chief difference between these regions' and Nova Scotia's climate was thus historical: the later arrival of able, hard-working settlers who could remake it into another temperate province of the British Empire.[1]

But learned elites' optimism about climate amelioration in settler colonies contrasted starkly with their disapproval of actual land-management practices. This is because their ideas about climate change cannot be adequately understood apart from their broader preoccupation with the aims, imperatives, and sharply critical discourse of agricultural improvement. Most improvers did not endorse the buoyant yeoman ideal that historians associate with Jeffersonian republicanism and its affirmation of self-sufficient small farmers. Hector St. John de Crèvecoer's well-known panegyric, *Letters from an American Farmer* (1782), was a rare exception. Like all improvers, Crèvecoer rehearsed a range of Enlightenment climatic, economic, and cultural theories to support the entitlements of colonial freeholders. The Lockean American farmer had taken a "formerly rude soil" and converted it "into a pleasant farm, and in return," he wrote, this labor "established all our rights; on it is founded our rank, our freedom, our power as citizens, our importance as inhabitants." British improvers ridiculed Crèvecoer's agricultural utopia as "a new species of forgery":

> The pen of this writer would make an Irish hut appear a palace most devoutly to be wished surrounded by a potatoe-garden, their cow flowing with healthful lacteal springs . . . In this manner the wretched inhabitants of the barren islands of Nantucket and Martha's Vineyard become the envy of those who enjoy every blessing which Nature kindly grants.

The sarcasm of this passage was typical of the language of improvement, which was often harshly critical in tone, particularly in its condemnation of land mismanagement in Britain and British America. As governing elites in Nova Scotia knew all too well, attracting loyal colonists of any stripe was a basic requirement for protecting British sovereignty, but without a population of diligent farmers, the colony would never be properly cultivated. "This province was first settled by vagrants and wretches who escaped the Law and were exempted from debt by fixing here, a method adopted to settle this part of America as being convenient for a Dockyard but which otherwise held very cheap," was the way Frances Wentworth candidly described the New England Planters in Nova Scotia. Only once enlightened governors like her husband secured more industrious settlers could they reliably develop the provincial economy—and, by extension, alleviate the province's climatic extremes—with the techniques of agricultural improvement.[2]

Improvement denoted a complex set of ideas and practices, offering a programmatic critique of traditional land-management practices and a set of new methods, materials, and technologies for agrarian reform that supposedly could be applied anywhere. First, it was forward looking. It rejected the circumscribed routines, antiquated means, and modest goals of village life in favor of a vision for "gradual, piecemeal, but cumulative betterment," in the historian Paul Slack's pithy phrase. Peasants, tenants, and smallholders were generally represented as a contemptible class, rustics who personified tradition ("culture derived from ancestors") and the negative effects of conservatism. Second, it was a program for modernizing and commercializing the economy. Improving landowners formulated standard recipes for increasing the yields, efficiency, and profitability of agricultural production. They recommended creating larger land enclosures; breeding bigger, more robust varieties of seed and livestock; experimenting with crop rotations and soil amendments; investing in mechanical equipment; and creating microclimates with forcing glasses, hothouses, and greenhouses. Third, it was a pragmatic response to Enlightenment theories of history, in which societies progressed, on different timescales, from primitive to civilized states.[3]

Beginning in the seventeenth century, writings about agriculture and natural history incorporating these ideas circulated throughout Europe and the Atlantic world in a range of public and private forms, adapted to a wide variety of environments and changing political contexts. Officials, landowners, and naturalists debated how best to implement improvement, casting and recasting it as a strategy for pursuing a range of competing or seemingly incompatible agendas: internal colonization at home and settler colonialism abroad; imperialism and anti-imperialism; patriotism and regionalism; cameralism and mercantilism; autarky and physiocracy; domestic self-sufficiency and commercial globalization. In this sense, improvement was an early modern concept as inclusive, versatile, and opaque as what is now called sustainable development. It was not only a set of formulas for increasing yields or land values but also a literary trope governed by rhetorical conventions and motivated by political interests. The discourse of improvement was so pervasive and protean that historians have often overlooked its diverse manifestations, including the idea that the climate itself could be subject to it.[4]

In theory, the improvement of people, lands, or climates transcended local conditions but in practice it was inevitably localized. Within Britain, improvers generally encouraged landlords to consolidate their estates, evict tenant farmers, and modernize the management of their properties using the latest scientific innovations. In North America, improvement was most concretely connected to the management of resources in plantations and settler colonies, but especially in the latter, improvers' prescriptions and predictions frequently extended into the intangible realm of the atmosphere. In New England and Nova Scotia, where local elites struggled in a variety of ways to overcome the real and imagined disadvantages of the northern environment, improvers encouraged industrious changes to the land that would ultimately temper the region's cold winters and humid summers. In the short term, agricultural improvement had to be modified to accommodate the topography, soils, and shorter growing season in the Northeast. By promoting scientific resource management across the region, however, improvers were also looking ahead to the future state of the Northeast. To determine whether colonial agriculture had already begun to effect historical changes to the climate, naturalists produced empirical reports describing, measuring, and analyzing wild environments, severe seasons, and temperature variations, a necessary first step in taking charge of improving the Northeast. They believed they were gradually creating mutual reforms in local society and the environment, hoping that settler landscapes and the skies above them would reify their inhabitants' growing cosmopolitanism. Progressive improvement would erode any sharp distinctions between the region and other cultivated, temperate places, promising to remake New England and Nova Scotia into true manifestations of their toponyms—that is, as re-creations or updates of the temperate British Isles—rather than their colder outposts.

The Energetic Hands of Britons

English landlords had long referred to profitable property reclamation or consolidation as improvement. In the early modern period, more ambitious rural land reforms, such as the enclosures imposed in regions of Scotland and Ireland, were promoted for their public virtues, from spurring economic development to raising the standard of living, whether introduced by private or government initiative. The English also considered the colonization of indigenous territories abroad as a form of improvement. Domesticating alien environments was a crucial step in making the margins of the world into more familiar and therefore seemingly more civil territory, as when New English planters secured a Protestant ascendancy in Ireland in the sixteenth and seventeenth centuries. In North America as in Ireland, the English pursued what historian David Armitage has called the agriculturalist justification for empire or "colonization through cultivation."[5]

While colonists praised American nature as abundant and fecund, they perceived it as virtually uncultivated, despite the fact that they benefited directly from Native American farmers' advice, seeds, provisions, and, ultimately, from the takeover of their fertile lands. Because the "whole earth is the lords Garden," explained John Winthrop, Sr., it was improvident to "suffer a whole Continent, as fruitfull & convenient for the use of man to lie . . . without any improvement." Believing that particular forms of land use reflected cultural achievements, colonial farmers could understand their work in transforming the indigenous landscape both as necessary for ensuring a "competence" for their families and as a Christian duty, proofs of the legitimacy of their presence on the land. A truly improved landscape was one marked by exclusive property boundaries and dominated by separate pastoral and arable fields. Settlers could prove their own "honesty and industry" by felling forests, enclosing and manuring their fields, erecting fences and outbuildings, and importing and enhancing seed and livestock. Even if early settlers made some "mistakes and errors . . . for the want of knowledge and experience in Agriculture," all migrants to North American colonies "first attempted to Improve this Country." Colonial improvement was "common sense and common information." They embraced improvement wherever they settled, casting the colonization of native territories as land reform.[6]

Interest in the agricultural potential of foreign lands was evident among all the expansionist early modern European states, but settler colonialism was particularly characteristic of the British in North America. Whether planting permanent settlements was a deliberate strategy of empire, colonists tended to emphasize the success of their agricultural settlements over military operations or religious crusades.[7] In the late seventeenth and eighteenth centuries, they employed the rhetoric of agricultural improvement to legitimate their conquest of French North America, justifying the crown's encroachment on Mi'kmaw and Acadian lands in strongly partisan, sometimes xenophobic, terms that figured farmers from the British Isles as superior in skill, vigor, and ambition. Many early British depictions of the peninsula's environment and French farmers were favorable to the extent that Acadian landscapes showed signs of agricultural potential, which British settlers would more fully exploit. Samuel Vetch invoked this contrast in "Canada Survey'd" when he wrote that only British inhabitants could turn the Acadian peninsula into a colony of agricultural settlement because French settlers would always "chuse rather to gain their bread by hunting and gunning, then by labouring the ground." In 1717, three British merchants wrote to the Board of Trade that, "the Soil of the Country is in General very good, abounding in the necessary Subsistence for Cattle, and bears all Sorts of European Grains in great Plenty where it is cultivated by the french." The lands were also "capable of great Improvements"—but this capability was not altogether realized under French possession. If the Board of Trade hoped to take advantage of these untapped riches they would have to send "thither a Colony of your Majesty's Subjects."[8]

When Acadian farmers refused to swear oaths of allegiance to British rule in Nova Scotia, officials initially formulated a policy of confiscating Acadian lands. In 1720, reports from British Nova Scotians to the Board of Trade again urged its cooperation in evicting French farmers and "the Indians who are their firm allies." Financing their dispossession and displacement would "prove effectual to establish the King's authority in this Province and facilitate the settling the same so as to prove in time advantageous to the Crown and to the Trade of Great Brittain." Paul Mascarene, a French Huguenot military engineer who accompanied Vetch from Boston to Nova Scotia, argued for the legitimacy of such seizures in terms of British settlers' ability to exploit the lands. He could tell from Acadian gardens that the soil was "rich in its produce," but that lands had "not improved as might be expected" because the French were "living in a manner from hand to mouth." He felt certain that Britons would make "better improvement of it for which their Industry is farr superiour to the french." Largely because no British settlers arrived in this period, the policy was not enforced.[9]

Much later in the century, when DesBarres was appointed lieutenant governor of Cape Breton, he wrote to the Board of Trade that, although British settlement on the island was currently sparse, he was encouraged by its strategic location for mediating the fisheries between the Gulf of Saint Lawrence and the North Atlantic, which "France valued most of all she ever held in North America" and had accrued "greater profits and more solid advantages than from all her other transatlantic territories in a century." He also "could not help pointing out" that Cape Breton contained "upwards of two millions of acres of Land the chief part of which is equally fit for culture as any in America—covered with all the species of useful timber common to the Provinces lying north of New York," but which the French had left uncultivated. "It appeared rational to anticipate," he assured the Board, that the lands would soon be improved "in the more energetic hands of Britons."[10]

Improvers' targets were never exclusively strangers or foreigners. Once colonists had secured territorial sovereignty and established a sufficient settler population in the region, improving elites increasingly began to emphasize the differences between "book" farmers like themselves and their untutored neighbors. Although officials expected all settlers to remake native, peculiar, and deficient environments into productive neo-European landscapes, at the same time they distinguished the practices and ambitions of ordinary farmers from their own conspicuously forward-looking, scientific, and cosmopolitan approaches. Although the Massachusetts Society for Promoting Agriculture solicited the experiential knowledge of so-called common or practical farmers and gave assurances that an inability to "write in a polished style" would not prevent them from publishing it, these overtures were largely attempts to win local support for and cooperation with the Society's improving mission, not to integrate local traditions into their schemes for agricultural reform.[11]

The tone of Arthur Young's strident editorials in a number of widely circulated agricultural publications expressed the key messages of improvement. Shortly after the independence of the United States, Young inaugurated his journal *Annals of Agriculture and Other Useful Arts* with a characteristically zealous statement of purpose. The *Annals* promoted scientific agriculture as "the plainest, most obvious" method of addressing the nation's war debt and of checking the expansion of empire. Celebrating the loss of the Thirteen Colonies, Young argued for further divestment from unnecessary overseas entanglements, which, so far, had enriched only a few "beggars, fanaticks, felons, and madmen of the kingdom" who "had been encouraged in their speculations of settling the wilds of North America." Colonization had drained the resources and stymied the development of the mother country. The countryside surrounding Cambridge, England, for example, was a scene of medieval torpor: "Bid the current of national improvement roll back three centuries," Young exclaimed, "and we may imagine a period of ignorance adequate" to describe this "beggarly village . . . such sloth—such ignorance—such backwardness—such determined resolution to stand still." While "every nerve strained to spread cultivation over American wastes, those of Britain have been left as if unworthy of all attention."[12]

Young's readers in North America agreed with the thrust, if not the particular examples, of his diatribes. As Young was urging Britons to focus on the problems of domestic agricultural underdevelopment and emigration to the colonies, North American landowners—in the remaining as well as former colonies—worried about regional underdevelopment and the problem of emigration to newly conquered lands in the Trans-Appalachian West. American farmers everywhere clung to "deeply rooted prejudices." An improver in Flushing, New York, and member of that state's agricultural society reiterated the standard complaint that "the Inhabitants of Long Island are not Celebrated for their Improvements in Agriculture—but here are a few who rise Superior to prejudice and the old beaten road of their forefathers & try new meathods of Improving their Lands & c." In a survey history of the eighteenth century, a fellow New Yorker made a more sweeping claim about American farmers in general:

> The improvements which have taken place in the agriculture of the United States, during the last twenty or thirty years, are very great. Our farmers, it is true, are far from having kept pace with their European brethren in enterprise, and the adoption of new and profitable modes of cultivation. Many of them obstinately adhere to practices which have been completely exploded; and neglect other and better, though recommended by the fullest experience. But if much remains to be done, much has also been performed towards the correction of this evil . . . Gentlemen of learning, observation, and property have zealously embarked in this interesting cause. The adoption of trans-atlantic improvements is gradually becoming more common.

From Quebec to Jamaica, criticisms of local farmers were specimens of the aggressively forward-looking discourse of improvement, which local elites translated to fit different environmental, economic, and political circumstances.[13]

In New England and Nova Scotia, improvers minimized the disadvantages of the local climate and soils while magnifying the bad practices of local farmers, blaming their backward-looking mentality and indifference to new, scientific agricultural methods. Jared Eliot's *Essays on Field Husbandry* repeatedly criticized the "indolence and carelessness" of Connecticut farmers. After reading the *Essays*, a prominent New Jersey man wrote to Eliot that "the Publick may be much benefited" by them, but "if the Farmers in your neighbourhood are as unwilling to leave the beaten road of their ancestors as they are near me, it will be difficult to persuade them to attempt any improvement." Later in the century, a Vermont man expressed "pity" that more farmers were not interested in reading the proceedings of the Massachusetts Society for Promoting Agriculture. "In the section of the Commonwealth where I reside," wrote a Cape Cod improver to the Agricultural Society, "most farmers are content to toil on in the same dull round that their Ancestors have done before them without troubling themselves much about alterations for the better." Likewise, Rhode Island's farmers allegedly practiced "only the traditionary husbandry of our fathers. What they introduced, we have continued. What it was a century ago, it is now. It has remained stationary at that point." Speaking of his visits with British improvers John Sinclair and Thomas William Coke, William Dandridge Peck reported to his superiors at Harvard that "prejudice is as strong in that country as in this, as any other, against the introduction of any improvement useful."[14]

Elites in Nova Scotia and neighboring islands also deployed the scornful rhetoric of improvement. On DesBarres's vast estates in northwestern Nova Scotia, John MacDonald described Acadian tenant farms as "some ridges of potatoes, some of barley, poor pease & oats." He saw this visual jumble as "a very unfarmerlike disposition of ground, as every plot ought to be prepared throughout for one & the same species of crop." MacDonald was a Highland laird turned Prince Edward Island landlord. DesBarres hired him as a surveyor to communicate with him about the state of his properties while he was in London. Relying on the standard imagery of improvement discourse, MacDonald wrote that the Acadians lived "settled or huddled together in form of a village." When he asked them why they traveled six miles round trip as a group on a daily basis to reach the coast where they grew marsh hay rather than cultivating fields closer to their homes, they responded that the uplands, or "caribou plains," were so "wretched that nothing could be made of them." But MacDonald dismissed this assessment and told them he "supposed the worst soil imaginable might be cultivated with good support." "I detest it," he sputtered in his report to DesBarres, referring to this exchange with Acadian farmers. "I hold it to be adverse to the progress of Improvement—and a nasty dirty way."[15]

MacDonald expressed the common opinion of improvers everywhere when he surmised that the Acadians were probably of equal capability with the New England Planters. That is, that both groups were equally incompetent: "true they are in several respects bad farmers & do not seem likely to improve; but the whole country are bad farmers and do not seem to do justice to the lands and if the Acadians are worse in some points than our sort are, they are better in others." One agent for the Board of Trade championed Acadians' ingenuity in developing dikes and hinged valves for regulating water flow in the tidal salt marshes across the Bay of Fundy coast and communicated his disapproval of absentee landlords who drove "poor unfortunate" Acadians from their tenancies, even though it was "a known fact that this Land previous" to French colonization was "as much in a state of Nature as it was when Noah came out of the Ark." Two Scots traveling through Nova Scotia in the 1770s described the Planters as "ignorant, indolent, bad managers," who had shown "neither the inclination nor industry to make great improvements." Newly arrived Loyalists frequently elevated Acadian farmers above the Planters. A British land agent who opposed another mass expulsion of Acadian families believed that the New England Planters were even more deserving of eviction. They had made only "slow progress" in cultivating Nova Scotia and could never improve or be improved because "ignorant of the true principles of husbandry and . . . full of Bigotry and Superstition, they disdained to avail themselves of Instruction." Another Loyalist settler told his correspondent "at the beginning of the American Revolution, this Province was just emerging from a state of wretchedness into which it had been plunged by the indolence of its Inhabitants and the bad Policy of its Government." He referred not to the "industrious Acadians," but to the Planters, whom he called "the refuse." For their part, Acadians adopted the language of improvement when it was instrumental in gaining sympathy and concessions from British landlords. One Acadian tenant farmer asked his landlord DesBarres for protection from escheat. He reminded "Mr. Dabar when you was here last you asured me that the land you settled me upon here was now belonging to you." If he were evicted, the man would be "disappointed to lose so much time and labour," so he begged DesBarres to allow him to "go on to Improve without fear."[16]

Although large landowners were rarely self-reflexive or ambivalent about their own abilities, MacDonald was sensitive to the fact that his Prince Edward Island estates also diverged from the British ideal of improvement. He admitted that "an observer from another country would certainly see much to be altered & corrected" in his field divisions, which ranged from one-quarter acre to four acres divided among many smallholders or tenants. British improvers were pointedly opposed to such small farms. They idealized instead large uniformly planted tracts or pastures, neatly separated by hedgerows or other permanent, rectilinear field divisions. These were some of the standard features of eighteenth-century enclosures and grand estates, especially in the English Midlands. The Board of Agriculture's surveys of the British countryside documented the prevalence of smallholdings and common pastures, overreliance on long-fallows, and outdated methods—evidence of mismanagement that Young, the Board's secretary, considered practically criminal. Small farmers, he exclaimed, were the "goths and vandals of the open fields." When a man identifying himself as a butcher in Grosvenor Market wrote to state his views on the "Evils of large Farms," the Board only agreed that farm size was "a subject on which opinions differ very materially, but as the Board wishes to hear both sides of the question, they have requested me to thank you for your letter." The patronizing tone of the reply indicated their position on the subject.[17]

Some large landowners in Nova Scotia won unconditional praise as agricultural improvers. The Loyalist Timothy Ruggles, for example, who was formerly speaker of the House of Representatives and surveyor general of the King's Forests in Massachusetts, resettled his family on a land grant of 10,000 acres in the township of Wilmot, Nova Scotia. A founder of the province's agricultural society, his fellow members admired that Ruggles, especially at such "a very advanced age," had "spared no expence in making experiments in Agriculture and Mechanics" and "revived" Wilmot, which had formerly looked like "Goldsmith's deserted village."[18] But the typical farms of the region rarely impressed visitors. Of a "farmer and landholder of some eminence" near Belfast, Maine, who planted only five of his eighty acres, La Rochefoucauld-Liancourt wrote "It is not easy to see, how old Nicholson can have acquired the reputation of being a good farmer." When James Skey, a Birmingham, England man, assessed the countryside in Connecticut, he expressed a typical newcomer's view: "This State is divided into small farms, and the little farmer, in this as well as in any other State is a poor miserable being." His fiancée Martha Russell, on the other hand, praised Connecticut, where her father had decided to settle. It seemed to her a tableau of model farms: "a fine undulating country, richly and extensively cultivated. Were I to sit down and in idea paint a country as I could wish to find it, no visionary fancies could approach nearer my wishes than this . . . reality." Maybe because most of the other places they visited on their New England tour fell so short of expectations, she reminded herself that "I ought to recollect that this country is more than as old again as some of [the other States], and has therefore had so much the longer time for improvement." Nevertheless, Russell declared Rhode Island, which was also settled in the early seventeenth century, to be entirely a "vile country" where farmers showed "apathy and indifference to everything" and particularly "to the gardens and farmyards." Russell and Skey intended to return to Birmingham. From what Skey had seen of southern New England, he decided that the only reason "a man should cross the Atlantic," was to appreciate that England was "infinitely more rich and bountiful than this continent."[19]

Improvers in northern climates agreed that scientific agriculture offered standard methods that promised to more fully integrate local farmers into broader commercial networks, but they also knew they could not rival warmer-climate plantations in the short term. As a result, they accommodated some of the fundamental aims of scientific agriculture to the necessarily smaller scale of operations in colder climates. In the Scottish Highlands, members of the Board of Annexed Estates pursued schemes for commercial mining and plantation agriculture by resettling tenant farmers on small crofts of one to two acres. The Russian tsar empowered naturalists and geographers to develop similar projects for resettling peasants and soldiers on small farms in the empire's steppe frontiers from the Ukraine to the Caucasus. Linnaeus's student Pehr Kalm traveled to other northern countries with a special interest in understanding the advantages of intensive agriculture on "the smallest pieces of land," which he believed was a laudable Chinese mode of farming.[20]

As in northern Europe, elites in New England and Nova Scotia were notably less antagonistic than English improvers to small-scale enterprise. For "gentleman of large estates, who can bear some considerable loss without feeling it," experimentation with "new crops, or new ways of raising old crops" was one legitimate form of small farming, especially if it involved investment in costly technologies for tinkering with microclimatic changes. When Prince Edward, the Duke of Kent, sent his gardener to oversee a botanical "depot" in Halifax, he envisioned building a hothouse for cultivating and then acclimatizing Caribbean plants to cold air until they could be shipped safely to Kew. Around the same time, Harvard installed a fuel-efficient English hothouse to keep tropical plants alive in its new botanical garden. The Vaughans' estate and commercial nursery in Maine included a greenhouse, "hot-beds," and piazzas to shelter delicate perennials; orchardists in Nova Scotia installed thermal lamps to prevent harvest failure from early frost. Homes with large windows and southern exposure could be kept bright and warm enough to grow oranges and pineapples through the winter, even if their flavor was "insipid."[21]

Beyond their gentrified estates, local elites perceived the social and environmental challenges to the practicability of extensive agriculture. In the American South and the West Indies, wrote Samuel Deane, "men choose to hold large farms, but in places where labour is dear, as in this country, small farms are to be preferred." In areas of sparse European settlement such as the District of Maine and Nova Scotia, improvers recommended enclosing newly claimed property in small sections. Sarah Vaughan, for example, decided the family needed to initially partition its grazing land "into two, if not into three parts, as small pastures answer best for small stocks of cattle." In the more densely settled older colonies like Connecticut, Jared Eliot promoted improvement on small tracts in part as a way to forestall westward migration among frustrated farmers who wanted to "have more Room, thinking that they live too thick." At the turn of the eighteenth century, improvers restated this point, insisting that since "the State of Connecticut has a comparatively thick population and that its territory is divided into small plantations or farms," landowners should be that much more reliant on natural history for understanding "how to enrich the soil they cultivate, and to extend its produce."[22]

Fieldwork

To better adapt scientific agriculture to the region, improving landowners and government officials sponsored natural history surveys that focused on the characteristics of the climate and the quality and quantity of local resources. Early American naturalists often frankly asserted that their prevailing interest in understanding local environments was to enhance their use and development. Edward Long stated as a truism that "whatever relates to climate and productions, is, to the naturalists, one of the chief objects of enquiry." Colonists simultaneously pursued scientific knowledge about the natural world and exploited it, assuming that natural history was not only compatible with but preliminary to resource development. In this sense, early American natural history was a speculative venture. Natural history surveys were organized not to merely satisfy curiosity but to commence official business such as mapping crown territory, determining provincial and private property boundaries, and appraising land values. Like early modern travel writing of any kind, surveys were composites in intent, practice, and record—various enough to exceed the genre. Private accounts of journeys through the region were recorded in published travelogues, personal journals, letters, and the minutes of society meetings. Official surveys describing the local climate, vegetation, and wildlife included hyperbolic promotional tracts aimed at attracting investors and settlers as well as more objective reports to the Royal Society or Board of Trade. The natural history expedition was a cousin to these personal travelogues as well as more formal demographic and statistical surveys, such as those that William Petty and John Sinclair initiated in Ireland and Scotland. As surveyor Samuel Holland wrote to the Board about

Cape Breton in 1766, "I shall endeavour in the Description of this Island, to give their Lordships what Intelligence I am Able, as I have made it my business to make myself acquainted with the former and present state of this Island and the manner of Improving it."[23]

Provincial practitioners especially tended to conflate natural history with agricultural improvement or justify the study of the former in the economic terms of the latter. They could point to Sir Joseph Banks—not only the president of the Royal Society but also founding member of the Board of Agriculture, the Royal Institution, and the Royal Horticultural Society—as a paragon of utilitarian natural history. Most early American naturalists believed, like Benjamin Waterhouse did, that utility and economy were "the ultimate ends" of science. Waterhouse declared that botany, chemistry, mineralogy, and zoology formed "the very basis of agriculture" and that, in turn, "every student of nature knows the dependence of agriculture on a correct natural history" (it was the Massachusetts Society for Promoting Agriculture, after all, that provided funding for Harvard's professorship of natural history and the college's botanical garden). Echoing this position, Nova Scotia improvers asserted that "great improvements in Natural History—particularly in Agriculture" were enhanced by the study of "Botany and Chemistry, both of which are subservient to Agriculture." Yoking natural history to economic development was "the best cure for local prejudices." It also made good business sense. "Commerce," William Peck bluntly told the Harvard Corporation, "is the friend of Science." Titus Smith expressed this relationship between research and development somewhat more critically: "The four quarters of the Globe are ransacked to supply man's wants, and he draws so much from the vegetable kingdom that it is necessary that some should be acquainted with a considerable share of its productions."[24]

Because the abiding goal of improvement shaped many forms of natural history writing, detailed descriptions of the climate or of wild plants and animals were often either embedded in agricultural improvement literature or natural histories described features of rural and urban landscapes. Gilbert White's *Natural History of Selborne*, which became a template for other local naturalists, began with a comparison of the fertility of the village's various soils. And because surveyors' descriptions of wild and cultivated nature were combined, they are sometimes difficult to differentiate in retrospect. Standard questionnaires or natural histories about particular areas presumed this continuum by asking about a range of seemingly heterogeneous information. For example, Belknap's third volume of his state history, subtitled "a Geographical Description of the State; with Sketches of its Natural History,

Productions, Improvements, and Present State," was compiled "from original surveys of many townships and tracts of the Country; from the conversation of many persons who have been employed in surveying, masting, hunting and scouting; as well as in husbandry, manufactures, merchandise, navigation and fishery." Under the heading "natural history," a circular issued in Prince Edward Island listed questions mainly asking about the potential for agricultural improvement: "Is there any Limestone? Is there any Plaster of Paris? Does it ever lay in large blocks? Is there any of a half transparent kind? Of what nature are the interior lands as to soil and situation; which are best calculated for present settlements? What are the wild animals?"[25]

Conversely, surveyors strained to couch descriptions of places that appeared disconcertingly barren in the language of colonial improvement. A Massachusetts man touring Annapolis Royal in the 1730s noted "there is no such thing as an oak, walnut or chestnut tree in these parts and the land so poor that no other trees grow to be above a foot or foot & half over, & very few so large." Despite the poverty of the area, he was reassured that agriculture was possible in Nova Scotia by the local provisions ("bonnyclabber, Soup, Sallet, roast shad & Bread & Butter & Roast mutton") his Acadian hosts fed him at dinner. Naturalists tended to omit details about areas they judged to be completely irremediable, implicitly reinforcing the idea that the study of particular environments was worthwhile only insofar as they could be exploited or improved. In another early report to the Board of Trade, an official in Nova Scotia summarized the landscape of French-controlled Île Royale (Cape Breton) in one sentence: "the Soil is no way valuable being intirely a Rock covered over with moss." Because surveyors had little incentive to provide detailed descriptions of environments they deemed too resource poor, ecological and economic geographies were subtly converged through omission. When Titus Smith surveyed Nova Scotia in 1801, he described the coast facing Cape Breton with the caveat that, because he was "informed by others who had traversed that part of the country that the land to the Westward of the road was chiefly barrens. . .we therefore concluded to shape our course accordingly." That is, they altogether avoided exploring the island's coast.[26]

The same combination of scientific inquiry and economic interests motivated expeditions into the White Mountains in New Hampshire and Maine, which were among the Northeast's most rugged and thinly inhabited areas. Although New Hampshire had been colonized in the early seventeenth century, there was little European settlement before the 1760s, and the mountains limited farming and commercial activity. To encourage more settlement, "a company of gentlemen," including Belknap, Cutler, and several members of the American Academy of Arts and Sciences hiked the mountains in the summer of 1784, using John Wentworth's 1774 survey as a guide. Most of the men were interested in the area's real estate potential. For Cutler, the tour was a scientific expedition during which he would record bird sightings, collect botanical specimens, and measure the height of Mount Washington—the highest peak north of the Carolinas and east of the Mississippi River—to provide Belknap with materials for revising his natural history of the state. A botanical study of the highest elevations in the region could corroborate or dispute leading theories about the climatic similitude of all alpine regions or the essentially unified biogeography of all northern countries across the globe. Barometric measurements could have contributed to an understanding of local weather.

When surveying New Hampshire's rocks and minerals, Belknap and Cutler were heartened to find flint and slate, which could be readily mined and commodified, and disappointed that "some specimens of rock chrystal have been found, but of no great value. No lime stone has yet been discovered." English explorers, beginning with Frobisher and Gilbert in the sixteenth century, failed to find mineral riches to justify the colonization of northern North America, turning instead to the proxy gold of fish, fur, and, their most durable conquest, land. Two centuries later, the hikers were studying mountain vegetation in part to assess possibilities for cultivating uplands. Indications of fertile soil in what was otherwise marginal terrain could seem to be evidence of latent improvement. Even if no valuable minerals were ever found other features of the high terrain, could bring "certain riches" to agricultural development in lowlands: the precipitation at the top of the mountain yielded "freshets, which bring down the soil, [to] the intervals below, and form a fine mould, producing by the aid of cultivation, corn and herbage in the most luxuriant plenty."[27]

The purpose of such field trips was thus threefold: The gentlemen toured marginal areas to contribute empirical data to the encyclopedic project of Enlightenment natural history, to perform their expertise by detecting known or new species, and, finally, to assess regional commercial possibilities. As it happened, however, the 1784 White Mountains survey contributed to these goals only "in part" because cold temperatures and fog got in their way. Belknap told members of the American Philosophical Society that the formidable "weather while we were in that region hindered us from making some observations which we intended." According to Cutler, "it happened, unfortunately, that thick clouds covered the mountains almost the whole time." The dense cloud cover "rendered useless" the sextant, telescope, barometer, thermometer, and other instruments he had packed. In addition, the cold air at higher altitudes "nearly deprived him of the use of his fingers." By the time the party reached the summit, all the gentlemen were numb-fingered: One of them tried to "engrav[e] the letters NH but was so chilled with the cold, that he gave the instruments to Col. Whipple, who finished the letters." Even on hot days much of mountainous New Hampshire disappointed them. Momentarily excited by "the appearance of a close-fed pasture" above the tree line, on closer inspection the green turned out "to be a mere mass of rocks, covered with a mat of long moss." The "extreme" temperatures and "barren plains" were especially bad for botanizing. "This is a most wretched place indeed," Cutler huffed in his diary, "miserable huts, on very poor, rocky, rough land, constantly uphill and down." Cutler's curiosity was all embracing, but in his travel notes he recorded the most satisfaction from spotting domesticated landscapes: a freshly mown lawn or a "picturesque" meadow. As a naturalist, Cutler valued careful and accurate reporting, but as an improver he desired mild weather and a prosperous agricultural scene.[28]

Modern Winters

In the 1780s, Samuel Williams surveyed the breadth of Massachusetts and the length of the Connecticut River Valley, from Long Island Sound to Lake Champlain, witnessing the full spectrum of improved and unimproved tracts in the region. It was from this comparative fieldwork in New England's "ancient cultivated parts" and its "inland and mountainous parts" that Williams derived his suspicion that the region's climate had been changing. As a Harvard professor of math and natural philosophy and—thanks to his friendship with Count Rumford—the only North American member of the Palatine Meteorological Society, Williams possessed a substantial collection of measuring instruments, which he used to record air, ground, and well-water temperatures. From these calculations, he noted "the heat decreases as we advance towards the north, in a country but little cultivated." This geographical contrast across the region was not just the result of latitude. It also mirrored the progression of southern New England's environment since the beginning of the colonial period—a freeze-frame of the diachronic process of colonial improvement. By the late eighteenth century, "the earth and the air in the cultivated parts of the country, are heated in consequence of their cultivation, ten or eleven degrees more, than they were in their uncultivated state" in the early seventeenth century. According to Williams, colonists had initiated the process that resulted in this 10° difference of temperature:

> In one century and a half this part of America has undergone a change which no country has ever passed thro' in so short a time. The forests near the Seacoasts have been cut down. The swamps have been drained. The face of the earth has been bid open to the influence of the wind and sun. And the wilderness has been changed into meadows, pastures, orchards, and fields of grain.

Forest removal exposed "the land to the full force of the solar rays"; planting meadows, pastures, orchards, and fields required diverting water from streams and swamps; increased numbers of settlers and their livestock filled the air with the hot "exhalations of thousands," especially in "large and populous cities" like Boston and New Haven. These "causes" resulted in a number of "effects": Tributaries and wetlands dried up, snowfall decreased, winds changed course. In addition to these disturbances, the most perceptible mutation was in the seasons, all of which were modified and moderated: "While our winters have been gradually becoming more short and mild the summer has become longer in duration but more cool and pleasant."

These changes were instances of an undeniable "general and universal" phenomenon. North America's warming temperatures were only the latest manifestation on a regional scale of an environmental *translatio imperii*, a much longer, global environmental process that had begun in ancient Palestine. "In every cultivated country of which we have any account the climate has been found to grow more mild and temperate, from the very first creation of the world until now." All inhabitable, cultivable regions were subject to the same "operation of causes and effects": Ancient amelioration and modern decline documented in the climate history of the eastern Mediterranean must have occurred in the western Mediterranean as well as in the populous regions of Asia, even though "we have not indeed any ancient, historical or philosophical accounts of Asia that will enable us to determine this matter from such kind of evidence."[29]

In the late seventeenth and eighteenth centuries, writers across the Atlantic world who contributed to debates about the theory of anthropogenic climate change collected evidence from a variety of biblical, classical, scientific, and historical sources. The idea that climate was an unstable entity subject to historical change was implicit in the Old Testament story of Noah and the Flood. But the further notion that human intervention could rehabilitate or ruin the climate was scattered throughout the work of ancient writers such as Theophrastus, Pliny, and Tacitus, in more recent philosophical tracts, colonial histories, and travelogues such as Swedish naturalist Pehr Kalm's frequently cited journey through the Middle Colonies and Quebec in the 1740s and 1750s, which introduced oral testimony as a source by reproducing conversations with elderly settlers about the weather during their childhoods. Those with access to measuring instruments also produced empirical evidence of weather and climate history, which they circulated in manuscript or published form. These reports included the date and measurement of annual first and last frosts, spring and fall harvests, animal migrations, daily instrumental records, descriptions of seasonable and unusual weather, earthquakes, volcanic eruptions, and astronomical sightings.[30]

In the northern colonies, descriptions of changes in the weather dated to the earliest settlements, where many expressed the belief that as the region was increasingly "cultured, planted, manured by men of industrie," as John Smith foresaw in 1616, positive changes to the northern environment would follow. In his journal for February 1631, John Winthrop referred to the impressions of colonists in neighboring Plymouth when he wrote, "It hath been observed ever since this bay was planted by Englishmen, viz. 7 years, that at this day the frost hath broke up every year" and "though we had many snows and sharp frosts, yet they continued not, neither were the waters frozen up as before." The narrative of climatic improvement was affirmed by many seventeenth-century writers who focused on the region, from Edward Johnson and William Hubbard in New England to Pierre Biard in Acadia and Jean Talon in Quebec. In the same decade, Philip Vincent expressed this hopeful view in his forecast of the environmental consequences of colonial victory in the Pequot War. Pointing out that Connecticut was located between the parallels of northern Spain and southern France, he predicted that its "temper" could also become "as pleasant, as temperate and as fertile as either, if managed by industrious hands." A contrarian example that indicates the idea's ubiquity—as well as its ideological adaptability—was a 1676 communication to the Royal Society asserting that the English conquest of Ireland had tempered the climate by inducing land abandonment and depopulation.[31]

By the eighteenth century, the idea of anthropogenic climate change—particularly the capacity of populous, industrious settler societies to improve local climates through profitable land management—gained adherents throughout the British Empire and the early United States. In the mid-eighteenth century, David Hume expressed the typical view that, though American climates were still colder than their European counterparts at the same latitude, "our Northern colonies in America become more temperate, in proportion as the woods are fell'd," an argument Philadelphia physician Hugh Williamson applied two decades later to the Middle Colonies. Even the author of *American Husbandry*, who dwelled at length on the northern colonies' "bad" climate, believed it had "been vastly improved since the country has been cleared of wood and brought into cultivation. The cold in winter is less intense, the air in summer purer, and the country in general much more wholesome." Providential changes were equally apparent to learned elites in the warmer climates of the Chesapeake and Lower South. As Thomas Jefferson wrote in his *Notes on the State of Virginia*, "a change in our climate . . . is taking place very sensibly. Both heats & colds are become much more moderate." And David Ramsay believed that "in the last seventy years" the "half-West-India climate" of South Carolina "had changed for the better." Ramsay resolved to continue measuring daily air temperature and barometric pressure, but he was nevertheless "certain that the climates of old countries have been materially improved by clearing and cultivating the land."[32]

To verify his own experimental results, Williams could integrate these reports with local weather observations recorded by earlier generations of colonists and quantitative information available in local publications or provided by his peers Ezra Stiles, Manasseh Cutler, and Edward Holyoke in southern New England, DesBarres in Nova Scotia, and Jean-Francois Gaultier in Quebec. Reading through historical descriptions of conditions in early colonial sources, he was struck by the consistent differences between past and contemporary experiences of the local climate. His chief example was the dates when Boston Harbor froze and thawed in the 1630s drawn from the traveler William Wood's *New England's Prospect*, which differed considerably from trends in the second half of the eighteenth century. Continuous improvements had shortened and lessened the severity of winter conditions in the vicinity of Boston, which, Williams believed, corroborated what Williamson, Kalm, and Gaultier had reported for Philadelphia, New York, and Quebec.[33]

In places with supposedly more severe climates and relatively small colonial populations, settlers looked forward to the prospect of tempering. The leaders of the Sierra Leone Company appealed to the standard theory in assuring British settlers worried about surviving tropical heat that the local climate "may reasonably be expected to improve in proportion as cultivation takes place." Moreover, climate change in recently colonized environments was supposed to be especially mercurial, so British West Africans expected fast results. As Williams explained, tempering was "most of all sensible and apparent in a new country, which is suddenly changing from a state of vast uncultivated wilderness, to that of numerous settlements, and extensive improvements." Accordingly, in the decade since Freetown had become the colony's capital, its climate was "believed to be already considerably ameliorated." Likewise, a surveyor who had first traveled to Nova Scotia with Vetch in the 1710s reported to the Board of Trade in 1730 that in "that part of ye Soil that has been cultivated," he found "the climate is mostly temperate." On the other hand, "no certain judgment can be made of the parts that have never been cleared." After twenty more years had passed, a promotional pamphlet for the province proclaimed that, though the climate was "not, indeed, so agreeable as in the southern Parts of France . . . it would certainly grow better and better every Day, in Proportion as the Woods are cut down, and the Country cleared and improved; and when the Country to the North of it comes to be a little inhabited and cleared, it may, perhaps, become one of the pleasantest Spots upon the Globe."[34]

There were colonial naturalists who regarded this notion of climate engineering as teleology or reckless fraud. "From 150 years experience" of American colonization, wrote John Mitchell in 1767, "neither the soil nor climate will admit of any such improvements." He was especially skeptical of the possibility of amelioration in the northern colonies, where "the length and severity of the winters, the late and backward springs, and shortness of the summer season, are unavoidable obstacles to all such improvements in agriculture." Quoting at length from Mitchell's deeply pessimistic account, Edward Long dismissed altogether the "vague and unfounded assertion" that northern North America's climate would be tempered by agriculture and industry. To ridicule optimists such as Lord Sheffield, who proposed in 1783 that Quebec and Nova Scotia could replace the Middle Colonies and Carolinas as the empire's breadbasket, Long satirized the popular reception of Joseph Priestley's theory about dephlogisticated air (that is, air relieved of a supposed combustible substance called phlogiston):

> We shall probably be told, that so far as regards the *physical* maladies of climate, we are to *hope* that in *process of time*, when the atmosphere of these regions shall be more impregnated with phlogistic particles from myriads of reeking dunghills, from the fumes of furnaces, from the fires and smoke of ten thousand crowded cities *hereafter to be built*, and by a general subjection of the soil to agriculture, carried on to the Arctic Circle, they may be considerably alleviated.

Long's impatience for the argument for climate improvement was influenced by his learned interest in Jamaica's climate and natural history, but this particular tract was also clearly motivated by his opposition to new mercantile restrictions on commerce between the British West Indies and the United States. Because he decided that even southern New England farmers were incapable of growing grain for export, he doubted that Nova Scotia and Quebec would be able to provision Jamaican plantations or that its small settler populations would absorb the former colonies' consumer demand for sugar. To insist otherwise, he sneered, was "not less absurd than if we were to talk of feeding the manufacturers and stocking the looms of Norwich and Manchester from the desarts of Iceland." Deforestation would only make British North America colder, less cultivable, and "as uninhabitable as Hudson's Bay." The outcome would be that Halifax Harbor and the Gulf of Saint Lawrence would remain icebound during Jamaica's hurricane season, a coincident climatic vulnerability proving the folly of an embargo on trade with American wheat and rice merchants. Long submitted that climate change could result only from a shift in the position of Earth's axis. Barring this highly improbable reorientation, Long agreed with Mitchell that there was "nothing to be done against nature."[35]

Others who accepted reports of milder weather in areas of European settlement contended that these observations could be explained only by physical causes inaccessible to human manipulation. They found a credible explanation of climatic change in the reverberating destruction caused in various places by strong geological activity during the eighteenth century: the 1755 Lisbon earthquake, the floating black particles emitted by the eruption of Iceland's Laki volcano in 1783, or the supposed rumbling of an obscure volcano in the Saint Lawrence River Valley, which a British physician stationed in Quebec believed accounted for the increasingly warmer temperatures he documented in his journal of daily thermometer readings as well as anecdotal information gleaned from the province's "oldest inhabitants."[36]

Building on secular historical accounts of long-term climate change that had developed over the seventeenth and eighteenth centuries, Buffon offered the most radically secular synthesis between geo-physical and anthropogenic explanations for climate change. In the seventeenth century, naturalists had begun to piece together geological evidence, including skeletal remains and fossils of extinct creatures found in Siberia and other far northern regions, which resembled living species in equatorial regions. The correspondence between modern and prehistoric animals located at such a striking geographical remove suggested that continents, oceans, and climates had undergone profound changes over the course of millennia, on a much larger timescale than the mere seven millennia that the biblical account allowed. In *Des époques de la nature*, Buffon expanded the chronology of Earth history to between 74,000 and 77,000 years divided into seven epochs marked by ecological revolutions. Since the beginning, Buffon argued, Earth's climate had been affected by the process of refrigeration (*refroidissement*). Originally very hot and uninhabitable like the sun, global cooling started in the North Pole, where creatures like the elephant and rhinoceros first emerged on the Earth's sole land mass. During the late fifth and sixth epochs, as glaciers advanced in the far north and continents separated, plants and animals were forced southward and dispersed across the relatively warmer latitudes in what became Europe, Africa, Asia, and the Americas. In the last several millennia, however, a paradox developed. Even as global cooling continued and glaciers continued to encroach on the habitable climates, most of the planet was mild enough to support human societies or other forms of life. The only explanation for why the world was not yet frozen was that in this seventh and final epoch, nature had become subject to the improving "power of humankind."

> The entire face of the earth today bears the imprint of the power of man, which, although subordinate to that of nature . . . is a wonderful assistant . . . it is with our hands that [the Earth] developed to its full extent, and she came by degrees to the point of perfection and magnificence which we see today.

Agricultural improvement would delay, if never entirely prevent, the inexorable process of cooling that would eventually engulf the whole Earth.[37]

Buffon's climate science was deeply influential if not universally accepted. Among those disputing "the favorite hypothesis of the celebrated Buffon" was John Leslie, a professor of mathematics and natural philosophy at the University of Edinburgh who performed experiments in the physics of heat, cold, and light and invented instruments related to this work, including a differential hygrometer and photometer. During the late eighteenth and early nineteenth centuries, Leslie applied insights from this laboratory research to developing a theory of global climate warming, because, as he stated: "The more I reflect on the proposition, that climates are growing gradually warmer, the more am I convinced of its reality." Although Leslie's work in early climate science is less well known, he detailed his revision of Buffon's global cooling theory in two texts: "On Heat and Climate" and *An Experimental Inquiry into the Nature and Propagation of Heat*. According to Leslie, the improved climates of the Northern Hemisphere were linked to melting glaciers. The "extremely slow" retreat of glacial ice in the southern regions of the Arctic and the Swiss Alps, where he had done fieldwork in 1796, were the reason that "the earth is growing continually warmer" or, at least "that the climate, over the whole of Europe, has assumed a milder character." He argued that "human industry"—"the clearing of the forests, the draining of the marshes"—might reduce seasonal extremes and create the sensation of a more temperate climate, but it had "no influence whatever in altering the average of temperature." However, because Leslie assumed that much of the "New World" was still covered in ancient forests and was affected by different wind patterns than the Eastern Hemisphere, he allowed the possibility that "America forms a curious exception to the general rule" of climate dynamics.[38]

What no scientist who examined climate history knew for certain—including those who accepted the possibility of anthropogenic causes—was whether colder, warmer, or anomalous seasonal conditions were an indication of superficial or permanent changes. Benjamin Franklin was open to "the Truth of the common Opinion, that the Winters in America are grown milder." "But whether enough of the Country is yet cleared to produce any sensible Effect," he wrote to Stiles in 1763, "may yet be a Question: And I think it would require a regular and steady Course of Observations on a Number of Winters in the different Parts of the Country you mention, to obtain full Satisfaction on the Point." Their mutual friend Benjamin Vaughan—who saw no evidence of climate change in the settlements of eastern Maine—was skeptical of climate change theory, particularly in regard to Williamson's attempt to apply it to understanding the etiology and geography of infectious diseases. Vaughan argued that Williamson's speculation that contagions decreased as the climate became more temperate was "useless unless we can shew why such changes should first be confined to Philadelphia, & then gradually visit other scattered spots along the United States, to the exclusion of others," where the landscape was equally improved.

Drawing on the same sources, Noah Webster was more resolute. He was completely unpersuaded by conjectures about a permanent warming trend. In his essay "On the Supposed Change in the Temperature of Winter," first read before the Connecticut Academy of Arts and Sciences in 1799, Webster countered the widespread prediction that warmer winter temperatures in "northern latitudes" would be everlasting. Webster attributed any "diminution" in severe weather to mere variability, singling out Williams for having "run into the error . . . of taking the accounts of a few *severe winters* as descriptions of *ordinary winters*." Timothy Dwight elaborated Webster's criticism, adding the suspicion that the region's erratic weather was the effect of timeless "revolutions of the atmosphere" rather than of radical changes to it. Bouts of severe and mild weather "in this country and probably of others in the Northern temperate zone," Dwight reasoned, were part of a natural distribution of "periods of hot and cold," which operated on timescales far beyond any generation's recollection or current scientific understanding in the eighteenth century.[39]

Nevertheless, beneath these disagreements there was fundamental consensus about the necessity, worth, and profound consequences of improvement. No one disputed the sentiment Williams expressed in his manuscript on climate history: "Thus does the goodness and benevolence of the Creator everywhere encourage and reward the diligence and industry of the human race. Not only the circumstance of individuals but the earth itself become improved by their industry and labour." While Webster and Dwight rejected "a material change" in the climate, they reinforced the view that colonization and cultivation had beneficially destabilized the indigenous nature of the Northeast. Webster concluded his essay by echoing most of Williams' observations. "The weather in modern winters," Webster declared "is more inconstant, than when the earth was covered with wood at the first settlement of Europeans in the country; that the warm weather of autumn extends further into the winter months, and the cold weather of winter and spring encroaches upon the summer; that the wind being more variable, snow is less permanent."[40]

Dwight confronted these uneven transformations while researching his book *Travels in New England and New York*. By documenting current conditions—from mineral deposits to local climatic conditions—he had initially hoped to depict the region's natural history "as it truly appeared, or would have appeared, eighty or one hundred years before." But what he found was a landscape in "different degrees of improvement" and weather as changeable as a cloud. Dwight observed that "the state of this country changes so fast, as to make a picture of it, drawn at a given period, an imperfect resemblance of what a traveler will find it to be after a moderate number of years have elapsed. The new settlements particularly, would in many instances scarcely be known, even from the most accurate description, after a very short lapse of time." As a result, assessments of the past and current patterns of the climate were a necessarily indeterminate guide for projecting its future states. Changes depended on fluctuations of population, the management of resources, and commercial development. Improvement was the common method and objective, but it could offer only an approximate idea of how and when its discrete stages would be apparent in any particular place. Depicting the countryside only "as it is in a state of nature, or as it will soon be in a state of complete cultivation"—that is, to represent only the beginning and end points of improvement—was an act of gross omission that ignored the dynamism of environmental change.[41]

Until such time when naturalists could account for this dynamism "with accuracy and precision," however, even Williams conceded that there was no definitive proof to conclusively substantiate or refute permanent climate warming. In the meantime, he joined others in encouraging locals to keep daily records of the weather so they could "deliver down to posterity the materials of information and knowledge that they may have it in their power to perfect in their day an account of what we can only begin in ours." If proper and thorough documentation marked the crucial difference between known and unknown climate changes, the lesson for naturalists was clear: It would be necessary to conduct further research and analyze it carefully. Like the ongoing project of improvement, the archive of climate history—and interpretations of it—would never be perfectly complete.[42]

Notes

In citing works in the notes, short titles have generally been used. Works frequently cited have been identified by the following abbreviations:

SOURCES AND INSTITUTIONS

AAAS	American Academy of Arts and Sciences
BL	Add. Mss. British Library, Additional Manuscripts
BW-HMS	Benjamin Waterhouse Papers, Countway Library, Harvard Medical School
DFP	Drowne Family Papers, John Hay Library, Brown University
DTC-BMNH	British Museum (Natural History), Sir Joseph Banks Correspondence
ESP	Ezra Stiles Papers, Beinecke Library, Yale University
EUL	Edinburgh University Library, Special Collections
HL	Houghton Library, Harvard University
HLA	Huntington Library and Archives
HNH-3	Jeremy Belknap, *History of New-Hampshire,* Vol. 3 (Boston, 1792)
JDF	Joseph Frederick W. DesBarres Papers, 1762–1894, MG23-F1, National Archives of Canada
KBG-BC	Kew Botanical Gardens, Library and Archives, Banks Collection of Original Manuscripts
LAC	Library and Archives of Canada
LJCMC	William Cutler and Julia Cutler, eds. *Life, Journals, and Correspondence of Reverend Manasseh Cutler, LLD, By His Grandchildren*, 2 vols. (Cincinnati: Robert Clarke & Co., 1888).
MHS	Massachusetts Historical Society
MSF	Morse Family Papers, Manuscripts, and Archives, Yale University Library

MSPA	Massachusetts Society for Promoting Agriculture
NCHV	Samuel Williams, *Natural and Civil History of Vermont* (1794)
NHH	New Hampshire Historical Society
NLS	National Library of Scotland, Edinburgh
NSA	Nova Scotia Archives
NSM	Nova Scotia Museum, Prescott Family Papers
NSSPA	Nova Scotia Society for Promoting Agriculture
PRO/CO	Colonial State Papers, Public Record Office, National Archives, Kew
RA GEO	Royal Archives, Windsor, Georgian Papers
RIA	Count Rumford Collection, Royal Institution Archives, London
RS	Royal Society
SR RASE, BI-IV	Royal Agricultural Society of England, The Museum of English Rural Life, The University of Reading
SVP-CU	Samuel Vetch Papers, 1708–1712, Rare Book and Manuscript Library, Columbia University
SWP-HA	Samuel Williams Papers, Harvard University Archives
SWP-HL	Samuel Williams Papers, Houghton Library
SWP-UV	Samuel Williams Papers, Special Collections, University of Vermont Library
WDPP	William Dandridge Peck Papers, Harvard University Archives
WMQ	*The William and Mary Quarterly*

INDIVIDUALS

BV	Benjamin Vaughan
BW	Benjamin Waterhouse
JB	Joseph Banks
JW	John Wentworth
SV	Sarah Vaughan
SW	Samuel Williams
WDP	William Dandridge Peck

INTRODUCTION

1. Samuel Williams to Joseph Banks, September 16,1789, Box 2, Folder 28, Samuel Williams Papers, Special Collections, University of Vermont Library (cf. BL Add. Mss. 8097.358); Samuel Williams, *The Natural and Civil History of Vermont* (Rutland, VT, 1794), 43–46 (New Haven to Burlington, migration and vegetation), 57 (cold decreases), 60–61 (ground temperatures); "Change of Climate in North America and Europe" (unpublished manuscript [ca. 1790]), Box 1, Folder 10, Samuel Williams Papers, 1752–1794, HUM 8, Harvard University Archives;

Samuel Williams Family Papers, Ms AM 2624, Houghton Library (Cambridge to Springfield); David C. Cassidy, "Meteorology in Mannheim: The Palatine Meteorological Society, 1780–1795," *Sudhoffs Archiv* 69(1) (1985), 24.

2. Hubert H. Lamb, *The Changing Climate: Selected Papers* (London: Methuen, 1966); H. C. Fritts and J. M. Lough, "An Estimate of Average Annual Temperature Variations for North America, 1602 to 1961," *Climatic Change* 7(2) (1985), 203–224; Daniel Houle, Jean-David Moore, and Jean Provencher, "Ice Bridges on the St. Lawrence River as an Index of Winter Severity from 1620 to 1910," *Journal of Climate* 20 (2007), 757–764; Shaun A. Marcott et al., "A Reconstruction of Regional and Global Temperature for the Past 11,300 Years," *Science* 339, 1198 (2013); John F. Richards, *The Unending Frontier: An Environmental History of the Early Modern World* (Berkeley: University of California Press, 2003), Chap. 1; Paul N. Edwards, *A Vast Machine: Computer Models, Climate Data, and the Politics of Global Warming* (Cambridge: MIT Press, 2010); Thomas M. Wickman, "Snowshoe Country: Indians, Colonists, and Winter Spaces of Power in the Northeast, 1620–1727" (unpublished PhD dissertation, Harvard University, 2012); Katherine A. Grandjean, "New World Tempests: Environment, Scarcity, and the Coming of the Pequot War," *William and Mary Quarterly* 68(1) (2011), 75–100; Geoffrey Parker, *Global Crisis: War, Climate Change, and Catastrophe in the Seventeenth Century* (New Haven, CT: Yale University Press, 2013), 445–456; Sam White, "The Real Little Ice Age," *Journal of Interdisciplinary History* 44 (2013), 327–352; William R. Baron, "1816 in Perspective: The View from the Northeastern United States," in C. R. Harrington, ed., *The Year Without a Summer? World Climate in 1816* (Ottawa: Canadian Museum of Nature, 1992), 124–144.

3. Gregory A. Zielinski and Barry D. Keim, *New England Weather, New England Climate* (Hanover, NH: University of New England Press, 2003), 231–256; William R. Baron, "Historical Climate Records from the Northeastern United States, 1640–1900," in Raymond S. Bradley and Philip D. Jones, eds., *Climate Since A.D. 1500* (New York: Routledge, 1995),74–91.

4. Benjamin Franklin to Ezra Stiles, May 29, 1763, in Labaree, ed.,*The Papers of Benjamin Franklin,* Vol. 10, 264–267; Hugh Williamson, "An Attempt to Account for the Change of Climate, Which has Been Observed in the Middle Colonies of North-America," *Transactions of the American Philosophical Society* 1 (Jan. 1, 1769–Jan. 1, 1771), 277–280; "Notes on the State of Virginia" (Manuscript, 1781–1785), Thomas Jefferson Papers, MHS.

5. Clarence J. Glacken, *Traces on the Rhodian Shore: Nature and Culture in Western Thought From Ancient Times to the Present* (Berkeley: University of California Press, 1967), Aristotle quoted on 94–95; James R. Fleming, *Historical Perspectives on Climate Change* (New York: Oxford University Press, 1998); Anthony Pagden, *Lords of All the World: Ideologies of Empire in Spain. Britain, and France, c. 1500–c. 1800* (New Haven, CT: Yale University Press, 1995).

6. Karen O. Kupperman, "The Puzzle of the American Climate in the Early Colonial Period," *American Historical Review* 87(5) (1982), 1262–1289.
7. On native horticulture and land use in the region before and after colonization, see William E. Doolittle, *Cultivated Landscapes of Native North America* (New York: Oxford University Press, 2000); Elizabeth S. Chilton, "The Origin and Spread of Maize (*Zea mays*) in New England," in John E. Staller et al., eds., *Histories of Maize: Multidisciplinary Approaches to the Prehistory, Linguistics, Biogeography, Domestication, and Evolution of Maize* (Boston: Elsevier Academic, 2006), 539–547; David Demeritt, "Agriculture, Climate, and Cultural Adaptation in the Prehistoric Northeast," *Archaeology of Eastern North America* 19 (Fall 1991), 183–202; William Cronon, *Changes in the Land: Indians, Colonists, and the Ecology of New England* (New York: Hill and Wang, 2003 [1983]).
8. Timothy Dwight, *Travels in New England and New York* (New Haven, CT, 1821), Vol. 1, 105.
9. Robert Boyle, *General Heads for the Natural History of a Country Great or Small Drawn Out for the Use of Travellers and Navigators* (London, 1691), 2–3; Jan Golinski, *British Weather and the Climate of Enlightenment* (Chicago: University of Chicago Press, 2007); Vladimir Jankovic, *Reading the Skies: A Cultural History of English Weather, 1650–1820* (Chicago: University of Chicago Press, 2000), Chap. 4. In the early nineteenth century, the British writer Luke Howard proposed Linnaean nomenclature for classifying cloud formations. See Luke Howard, *On the Modification of Clouds*, 3rd ed. (London, 1864 [1803]).
10. George-Louis Leclerc, comte de Buffon, *Histoire Naturelle: générale et particulière. Des époques de la nature*, vol. 5 (Paris, 1778), 1 ("*dans l'histoire naturelle, il faut fouiller les archives du monde*"); Rhoda Rappaport, *When Geologists Were Historians, 1665–1750* (Ithaca, NY: Cornell University Press, 1997); Martin J. S. Rudwick, *Bursting the Limits of Time: The Reconstruction of Geohistory in the Age of Revolution* (Chicago: University of Chicago Press, 2005). The Anglophone scholarship on Buffon has focused mainly on his remarks about the degenerative effects of the New World's climates. See Antonello Gerbi, *The Dispute of the New World: The History of a Polemic, 1750–1900*, Rev. ed. translated by Jeremy Moyle (Pittsburgh, PA: University of Pittsburgh Press, 1973); Jorge Cañizares-Esguerra, *How to Write the History of the New World: Histories, Epistomologies, and Identities in the Eighteenth-Century Atlantic World* (Palo Alto, CA: Stanford University Press, 2001).
11. Michael A. Osborne, "Acclimatizing the World: A History of the Paradigmatic Colonial Science," in Roy M. MacLeod, ed., "Nature and Empire: Science and the Colonial Enterprise," *Osiris*, 2d ser., 15 (2000), 136–139; Joyce E. Chaplin, *Subject Matter: Technology, the Body, and Science on the Anglo-American Frontier, 1500–1676* (Cambridge: Harvard University Press, 2001), Chapter 4.
12. William Bradford, *History of Plymouth Plantation,1606–1646*, William T. Davis, ed. (New York: Scribner, 1908), 55 (small discontentments), 96 (hideous);

Philip Vincent, *A True Relation of the Late Battle Fought between the English and the Pequot Savages . . . With the Present State of Things There* (London, 1638), 2–3; John Winthrop, *Reasons to be Considered for Justifying the Undertakers of the Intended Plantation in New England* (London,1629); David D. Hall, *Worlds of Wonder, Days of Judgment: Popular Religious Belief in Early New England* (New York: Knopf, 1989), 166–167; Stewart H. Holbrook, *The Yankee Exodus: An Account of Migration from New England* (New York: Macmillan, 1950), 1–3.

13. Max Weber, *The Protestant Ethic and the Spirit of Capitalism* (New York: Routledge, 2001), esp. Chap. 5; Perry Miller, *Errand Into the Wilderness* (Cambridge: Harvard University Press, 1984 [1956]); Immanuel Wallerstein, *The Modern World-System II: Mercantilism and the Consolidation of the European World Economy, 1600–1750* (Berkeley, CA: University of California Press, 2011 [1980], 179.
14. For the regional variant of what Jan de Vries has called the moral peasant/acquisitive peasant debate, see Richard L. Bushman, "Markets and Composite Farms in Early America," *William and Mary Quarterly* 55 (1998), 351–374; Darrett B. Rutman, "Governor Winthrop's Garden Crop: The Significance of Agriculture in the Early Commerce of Massachusetts Bay," *William and Mary Quarterly* 20(3) (1963), 396–415; James Henretta, "Families and Farms: Mentalité in Early America," *William and Mary Quarterly* 35 (1978), 3–32; Winifred Rothenberg, *From Market-Places to a Market Economy: The Transformation of Rural Massachusetts, 1750–1850* (Chicago: University of Chicago Press, 1992); Allan Kulikoff, "Households and Markets: Towards a New Synthesis of American Agrarian History," *William and Mary Quarterly* 50 (1993), 340–355; Christine Heyrman, *Commerce and Culture: The Maritime Communities of Colonial Massachusetts, 1690–1750* (New York: Norton, 1984); Stephen Innes, *Creating the Commonwealth: The Economic Culture of Puritan New England* (New York: Norton, 1995). For Nova Scotia, see Beatrice Craig, *Backwoods Consumers and Homespun Capitalists: The Rise of Market Culture in Eastern Canada* (Toronto: University of Toronto Press, 2009); Danny Samson, *The Spirit of Industry and Improvement: Liberal Government and Rural-Industrial Society, Nova Scotia, 1790–1862* (Montreal: McGill-Queen's University Press, 2008).
15. Cronon, *Changes in the Land*; Ted Steinberg, *Nature Incorporated: Industrialization and the Waters of New England* (New York: Cambridge University Press, 1991); Carolyn Merchant, *Ecological Revolutions: Nature, Gender, and Science in New England* (Chapel Hill: University of North Carolina Press, 1989); Brian Donahue, *The Great Meadow: Farmers and the Land in Colonial Concord* (New Haven, CT: Yale University Press, 2004).
16. Alfred W. Crosby, *Ecological Imperialism: The Biological Expansion of Europe, 900–1900* (New York: Cambridge University Press, 1986); Thomas Dunlap, *Nature and the English Diaspora: Environment and History in the United States, Canada, Australia, and New Zealand* (New York: Cambridge University Press, 1999); Mark Harrison, *Climates and Constitutions: Health, Race, Environment*

and British Imperialism in India (New York: Oxford University Press, 1999); J. R. McNeill, *Mosquito Empires: Ecology and War in the Greater Caribbean, 1620–1914* (New York: Cambridge University Press, 2010). Besides the recent swell of studies placing early American history in the broader chronology of the Little Ice Age, exceptions to these generalizations include Kupperman, "Puzzle of the American Climate"; "Chaplin, *Subject Matter*; Wickman, "Snowshoe Country"; and Michael R. Hill, "Temperateness, Temperance, and the Tropics: Climate and Morality in the English Atlantic World, 1555–1705," (unpublished PhD dissertation, Georgetown University, 2013).

17. "Nicholson and the Council of War at Annapolis Royal to the Queen," [?] Oct. 1710, PRO, CO 5:9; Dwight, *Travels in New England and New York*, Vol. 1, 65 (northern temperate zone).
18. On the history of agricultural improvement, see Harriet Ritvo, *Animal Estate: The English and Other Creatures in the Victorian Age* (Cambridge, MA: Harvard University Press, 1987); Richard Drayton, *Nature's Government: Science, Imperial Britain and the Improvement of the World* (New Haven, CT: Yale University Press, 2000); Fredrik Albritton Jonsson, *Enlightenment's Frontier: The Scottish Highlands and the Origins of Environmentalism* (New Haven, CT: Yale University Press, 2013).
19. Richard White, *"It's your Misfortune and None of My Own": A History of the American West* (Norman: University of Oklahoma Press, 1991), 150–154, 196; Robert Grant Haliburton, *The Men of the North and Their Place in History* (Montreal, 1869), 2; Vilhjalmur Stefansson, *The Northward Course of Empire* (New York: Harcourt, 1922); Agnes C. Laut, *Canada: The Empire of the North* (Boston: Ginn and Company, 1909); Charles Pickering, *The Races of Man and Their Geographical Distribution* (Philadelphia, 1848); Mart A. Stewart, "'Let Us Begin with the Weather': Climate, Race, and Cultural Distinctiveness in the American South," in Mikulas Teich, Roy Porter, and Bo Gustafsson, eds. *Nature and Society in Historical Context* (New York: Cambridge University Press, 1997), 240–256; Warwick Anderson, *The Cultivation of Whiteness: Science, Health, and Racial Destiny in Australia* (Melbourne: Melbourne University Press, 2005); Alison Bashford and Carolyn Strange, *Griffith Taylor: Visionary Environmentalist Explorer* (Canberra: National Library of Australia, 2008).
20. Naomi Oreskes, "The Scientific Consensus on Climate Change," *Science* 306 (December 2004), 1686; Jane McAdam, *Climate Change, Forced Migration, and International Law* (New York: Oxford University Press, 2012); Candis Callison, *How Climate Change Comes to Matter: The Communal Life of Facts* (Durham, NC: Duke University Press, 2014).
21. Dipesh Chakrabarty, "The Climate of History: Four Theses," *Critical Inquiry* 35 (Winter 2009), 197–222; idem, "Climate and Capital: On Conjoined Histories," *Critical Inquiry* 41(1) (Autumn 2014), 1–23; cf. Fabien Locher and Jean-Baptiste Fressoz, "Modernity's Frail Climate: A Climate History of Environmental Reflexivity," *Critical Inquiry* 38(3) (Spring 2012), 579–598.

CHAPTER 1

1. Morton, *New English Canaan or New Canaan* (Amsterdam), 11–16, 94; Bradford, *History of Plymouth*, 65; "Manuscript notebook of Dr. William James Almon, c. 1780-1800," Almon Family Papers, MG1, NSA.
2. Kupperman, "Puzzle of the American Climate," 1262–1289; "Fear of Hot Climates in the Anglo-American Colonial Experience," 213–240; "Climate and Mastery of the Wilderness," 3–38; Jan Golinski, "American Climate and the Civilization of Nature," in Nicholas Dew and James Delbourgo, eds., *Science and Empire*, 153–174.
3. Quinn and Quinn, eds. *English New England Voyages*, 159; Bourne, ed. *The Voyages of Samuel de Champlain*, 4; Biard, *Relation de la Nouvelle France, de ses terres, naturel du pais, & de ses habitants*, 9; John Smith, *A Description of New England*, 10, 14; Alexander, *An Encouragement to Colonies*, 40; Wood, *New England's Prospect*, 3; Morton, *New English Canaan*, 11–12, 15, 94.
4. John Smith, *A Description of New England*, 10; Bradford, *History of Plymouth*, 95 (mentions Smith), 96, 100; [William Bradford], *Beginning and Proceedings of the English Plantation*, 16 ("iron"); Dunn and Yeandle, eds., *Journal of John Winthrop*, 28 (strawberries), 31–35; Hutchinson, *History of Massachuset's Bay*, 20–23 (hovels), 483–484 (strawberry). On freezing deaths among English and Indians in southern New England, see Wickman, "Snowshoe Country," Chap. 1.
5. On eighteenth-century images of northern Europe, see Oslund, "Imagining Iceland," 313–334; and Jonsson, *Enlightenment's Frontier*, Chapter 2. On Enlightenment biogeography, see Withers, *Placing the Enlightenment*; Browne, "Biogeography and Empire," in Jardine, Secord, and Spary, eds., *Cultures of Natural History*, 305–321; and Browne, *The Secular Ark*, 1–31.
6. Morton, *New English Canaan*, 94; Stewart, "'Let Us Begin With the Weather': Climate, Race, and Cultural Distinctiveness in the American South," in Teich, Porter, and Gustafsson, eds. *Nature and Society in Historical Context*, 240–256; Young, *The Farmer's Letters*, 473–474 ("diametrically," "grave"). Young was referring to accounts of the Floridas but in the service of making a general point about all descriptions of North American regions.
7. Benton, *A Search for Sovereignty*, 1–4; "Legal Spaces of Empire," 700–724; and "Spatial Geographies of Empire," 19–34; see also Mancke, "Spaces of Power in the Early Modern Northeast," in Hornsby and Reid, eds., *New England and the Maritime Provinces*, 32–49.
8. On the instability of early American borders, see Ayers and Onuf, eds., *All Over the Map*; Chaplin, *Subject Matter*, 152–153; Edelson, *Plantation Enterprise*, 18; Vetter, "Wallace's *Other* Line," 89–123; Zilberstein, "Natural History of Early Northeastern America," in Jørgensen, Jørgensen, and Pritchard, eds., *New Natures*.
9. Oldmixon, *The British Empire in America*, 7; Moll, *A System of Geography*.

10. Peter Browne, *The Civil and Natural History of* Jamaica, xxxii; Jeremy Belknap to Jedediah Morse, July 26, 1786, Folder 1: 3A, 1786–June 1789, Correspondence, Jeremy Belknap Papers, New Hampshire Historical Society; Williams, *Natural and Civil History of Vermont*, 46.
11. "climate, n.1. a-b" and "oecumene | ecumene, n.," *OED Online*, 2013. (Oxford University Press,accessed June 26, 2013); Dilke, *Greek & Roman Maps*.
12. Best, *Three Voyages of Martin Frobisher*, 4–6; Robertson, *The History of America: Books IX and X*, 56–57 (far to the north).
13. Biggar, *The Voyages of Jacques Cartier*, 35–36. For the late eighteenth-century echo of this corrective, see William Robertson, cited in Belknap, *A Discourse*, 75.
14. Bodin, *Method for the Easy Comprehension of History*, 85–90; Claude D'Abbeville, *Histoire des pères Capucins*,192 ("Je n'estime pas . . . qu'il y aye lieu plus temperé et plus delicieux que ce païs là"); Best, *Three Voyages of Martin Frobisher*, 3, 10–11; Chaplin, *Subject Matter*, 43–55; Mancall, *Fatal Journey*, Chap. 4; and idem, "The Raw and the Cold," 3–40.
15. Moxon, *Mathematicks Made Easie*, 181; Headley, "Sixteenth-Century Venetian Celebration of the Earth's Total Habitability," 1–27; Gómez, *Tropics of Empire*; Safier, "Transformations de la zone torride," 143–172.
16. Best, *Three Voyages of Martin Frobisher*, 10–11.
17. Quinn and Quinn, eds., *English New England Voyages*, 209. On the Little Ice Age in the Northeast, see David Anderson, "Climate and Culture Change," 143–86. On colonial responses to colder, stormier conditions in the early seventeenth century, see Grandjean, "New World Tempests," 75–100.
18. John Smith, *Generall Historie of Virginia*, quoted in Quinn and Quinn, eds., *English New England Voyages*, 353–354; Winslow, *Good Newes From New England*, 12, 22; Bradford, *History of Plymouth*, 96.
19. November 2 and 30, 1749, Ebenezer Parkman Papers, 1718–1789; March 18, 1749, Samuel Chandler Diary 1746–1772, Massachusetts Historical Society (hereafter MHS); Solomon Drowne to Sally Drowne, 20 March 1799, Series 4, Box 23, The Drowne Family Papers, Ms. Drowne, Brown University Library.
20. "Letter concerning Acadie and N.S." Samuel Vetch to the Lords of Trade, November 24, 1714, MG 1, v. 1520, Folder J, NSA; For icicles, see quotations from Charlevoix in Jefferys, *Natural and Civil History of the French Dominions*,1–2, and Mitchell, *Present State of Great Britain and North America*, 169; "The History of the Province of Nova Scotia," 27, (unpublished manuscript, 1790), Papers of Professor Andrew Brown (1763–1834), Edinburgh University Library, Special Collections. For colonists' experiences of the particularly severe winters of the 1620s, see Grandjean, "New World Tempests," 75–100.
21. Voyage of Sir Humphrey Gilbert (1583), quoted in Burrage, *Early English and French Voyages*, 188 (North and colde); Davis, *Seaman's Secrets*, n.p. [part 2] ("temperate arctic zone"); Daniel Neal, *History of New England*, Vol. 2, 564; June 30, 1774, Luke Harrison to William Harrison, MG1, Vol. 427, no. 187, NSA;

January 21, 1778, Blagden to JB, 167–169; March 26, 1778, 187, Sir Joseph Banks Correspondence, DTC, British Museum (Natural History) [DTC-BMNH]; "The situation of the University at Cambridge and state of the Country round it," Box 1 f. 52, Samuel Williams Papers, Special Collections, University of Vermont Library.

22. Hubbard, *General History of New England*, 19–21; Neal, *History of New England*, Vol. 2, 564; Dwight, *Travels in New England and New York*, Vol. 1, 58. On the transatlantic asymmetry of climates at the same latitude, see also Arbuthnot, *Concerning the Effects of the Air*, 77–78.
23. John Mitchell, *Present State of Great Britain and North America*, Vol. 1, 167 ("much worse"), 171 ("fisheries"), 257–260 ("deceived"); Wynne, *General History of the British Empire in North America*, 411–412.
24. Anon., *American Husbandry: Containing an Account of the Soil, Climate, Production, and Agriculture* (London: J. Bew, 1775), 1–3, 13–14, 49–50. Paraphrased in Franklin, *Philosophical and Political History*, 121. On the authorship of *American Husbandry*, see Carrier, "Dr. John Mitchell," 201–219; Carman, ed., *American Husbandry*; and Davis, "Richard Oswald as 'An American'," 19–34.
25. Quinn and Quinn, eds., *English New England Voyages*, 159, 209–210, 222; John Smith, *A Description of New England*, 10 (gardens); Thomas, "Contrastive Subsistence Strategies and Land Use," 1–18; Peace, "Deconstructing the Sauvage / Savage," 1–20.
26. Morse, *History of America*, Vol. 1, 18; Papers of Professor Andrew Brown, 21; Montesquieu, *Spirit of the Laws*, Vol. 1, 319, 385.
27. Gillies, ed., *Aristotle's Ethics and Politics*, Vol. 2, 227; Clifton, trans., *Hippocrates Upon Air, Water, and Situation*, 27; Buffon, *Natural History of the Horse*, 163.
28. Wolff, *Inventing Eastern Europe*, 4–5, 318; Davidson, *The Idea of the North*; Gómez, *The Tropics of Empire*.
29. Gillies, *Shakespeare and the Geography of Difference*; Scodel, *Excess and the Mean*, 1 (vicious); Floyd-Wilson, *English Ethnicity and Race*; Snider, "Hard Frost: 1684," 8–32; Melzer, *Colonizer or Colonized*.
30. Gibbon, *Decline and Fall of the Roman Empire*, 27–28; Montesquieu, *Spirit of the Laws*, Vol. 1, 378, 380, 385.
31. Buffon, *Natural History, General and Particular*, 127–129, 134; idem., *Natural History of the Horse*, 163–164.
32. Goldsmith, *History of the Earth*, Vol. 2, 6; Robertson, *History of America*, Vol. 2, 282.
33. Samuel Williams Family Papers, Series: II. Compositions and documents #67, #77, "Notes on the State of Vermont Began Jan. 27. 1792," Houghton Library; Dwight, *Travels in New England and New York*, Vol. 1, 18–19, 65. For the standard overview of the debate over New World degeneracy, see Gerbi, *Dispute of the New World*.
34. [n.a.], *A True Account of the Colonies of Nova Scotia and Georgia* (Newcastle: Printed in the present year [1780?]), 11, 21 (colonies); Samuel Vetch, "Canada Survey'd, or the French Dominions upon the Continent of America briefly considered,

(July 27, 1708)," CO 323/6, Nos. 64, Colonial State Papers, National Archives, Kew (Eastern); William Sabatier, "Circular on behalf of the compilers of a history of Prince Edward Island," Lt. Governor PEI, 1804–1812, Series 5, Joseph Frederick W. DesBarres Papers, 1762–1894, National Archives of Canada (climates); Benjamin Vaughan to Judge Lowell, March 19, 1800, Folder: Misc. Agricultural Papers, Carton 22, Vaughan Family Papers, Correspondence, 1773–1812, MHS (states); October 1789, "Proceedings of the Privy Council relative to the Hessian Fly," *Nova Scotia Magazine* (the more northern); Volney, *View of the Soil and Climate*; March 30, 1790, William D. Peck to Jeremy Belknap, Box 1 HUG 1677, WDPP (this country).

35. [Notes on botany], n.d., Folder 5, item 24, Titus Smith Jr. Papers, MG1, NSA (meteorological predictions); "Journal of Titus Smith," RG1 Vol. 380b, NSA (number); Cook, *Indian Population of New England*, Vol. 12; Harley, "New England Cartography," 169–196.
36. Quinn and Quinn, eds., *English New England Voyages*, 3 ('seductive'); Hutchinson, *History of Massachuset's Bay*, 5 ('unacquaintedness'); Gorges, *Advancement of Plantations*, 51 ("sea to sea"); Alexander, *An Encouragement to Colonies*, 34–35; Mancke, "Spaces of Power in the Early Modern Northeast," in Hornsby and Reid, eds., *New England and the Maritime Provinces*, fn. 21, pp. 332–333; Reid, *Essays on Northeastern North America*, 23–39; Taylor, *Liberty Men and Great Proprietors*, 12–14; Hannah Farber, "Rise and Fall of the Province of Lygonia," 490–513; D'Abate, "On the Meaning of a Name," in Baker et al., eds., *American Beginnings*, 61–88; Edney, "The Irony of Imperial Mapping," in Akerman, ed., *The Imperial Map*, 28–29. On the cartographical history of Massachusetts Bay Colony's expansionist aims, see Edney, "Printed But Not Published: Limited-Circulation Maps of Territorial Disputes in Eighteenth-century New England," in Schip and van der Krogt, eds. *Mappae antiqua*, 147–158.
37. Josselyn, *New England's Rarities*, 2, 4–5; Reid, *Essays on Northeastern North America*, 34; Pedley, "Map Wars," 96–104. On the role of maps in border and property disputes in New York, see Gronim, "Geography and Persuasion," 373–402. On early seventeenth-century English attempts to assimilate New Netherlands into maps of New England, see Schmidt, "Mapping An Empire," 549–578.
38. Mascarene, Description of Nova Scotia, 1720/21, MG1, v. 1520, Folder P, Mascarene (Engineer), NSA; Neal, *History of New England*, Vol. 2, 563.
39. Mitchell, "Sheet 2: Hudson Bay and north central North America," in *Map of the British and French Dominions*; Burke, *European Settlements in America*, Vol. 2, 129.
40. Morison, ed., *William Bradford's Of Plymouth Plantation,* quote on 76; Robert Rogers, *A Concise Account of North America*, 101 ("general name"); Douglass, *A Summary Historical and Political,* Vol. 1, 7; Mitchell, *Present State of Great Britain and North America*, Vol. 1, 167–168.
41. Young, *Annals of Agriculture*, Vol. 1, 1 (1784), 13; Klockhoff, "A Chorographical Map,"; Hornsby, *Surveyors of Empire*, 39, 41–43 (Potomac River); Pedley, *The Commerce of Cartography*, 123–126 (Northern and Southern Districts).

42. Jedidiah Morse to Jeremy Belknap, January 18, 1788, Correspondence, Box 1, Folder 10, Morse Family Papers (MS 358). Manuscripts and Archives, Yale University Library; Morse, *The American Geography*, 2nd ed., iv, vi, 140, 193; McCorkle, *New England in Early Printed Maps,* 114; Demeritt, "Representing the 'True' St. Croix," 516.
43. Belknap, *History of New-Hampshire,* Vol. 3; Williams, *Natural and Civil History of Vermont*; Also Sullivan, *History of the District of Maine.*The major American model for these works was Thomas Jefferson's *Notes on the State of Virginia* (1781–1784).
44. On the practice of chorography and its ideological aspects, see Withers, "Reporting, Mapping, Trusting," 497–521; Cormack, " 'Good Fences Make Good Neighbors,' " 639–661; and Richard Helgerson, "The Land Speaks," 50–85.
45. David Ramsay to Jedidiah Morse, November 30, 1787, in Brunhouse, "David Ramsay," 1–250, letter on 116–117.
46. Morse, *American Geography,* 8.
47. Morse, *History of America*, Vol. 1, 2–3; Dwight, *Travels in New England*, Vol. 4,233; Volney, *View of the Soil and Climate,*1–2, 6–7; Sahlins, *Boundaries.*
48. Pennant, *Arctic Zoology* 1st ed., Vol. 2, Supplement, 3.
49. June 20, 1780, MC to Professor Samuel Williams in Cutler and Cutler, eds., *Life, Journals, and Correspondence,* 80–82.
50. Cutler, "An Account of Some of the Vegetable Productions," 1 (1783), 396 ("climate"), 401 ("admit"); Bigelow, *Florula bostoniensis*, vi-vii.
51. Nov. 21, 1715, Thomas Caulfield to Board of Trade, Folder K, MG1 v. 1520, NSA; Belknap, *History of New-Hampshire*, Vol. 3, 120; "Notes on the State of Vermont," Samuel Williams Family Papers, Ms AM 2624, Houghton Library; Li, "Luigi Castiglioni," 51–56.
52. Deane, *The New-England Farmer,* 2–3, 58 (Dickson), 152 ("naturalist . . . latitude"), 83, 108, 115, 175, 184, 287 ("our climate"); Dickson, *A Treatise of Agriculture*, Vol. 2, 5–6; Johnson, "Climate," in *A Dictionary of the English Language*, 2nd ed., Vol. 1 [n.p.].
53. Burnett, "Hydrographic Discipline Among the Navigators: Charting an "Empire of Commerce and Science in the Nineteenth-Century Pacific," in Akerman, *The Imperial Map*, 247; Benton, *A Search for Sovereignty.*
54. Wood, *New England's Prospect*, 4, 7; Mitchell, *Present State of Great Britain and North America*, Vol. 1, 169.
55. James, *Three Visitors to Early Plymouth*, 17 (thick fog); Moxon, *A Brief Discourse of a Passage*, 3; Boyle, *General History of the Air*, 161–162; Zielinski and Keim, *New England Weather,* 61–63; Neal, *History of New England*, Vol. 2, 563–564 (vast lakes); Belknap, *History of New-Hampshire,* Vol. 3, 17–18 (vulgar; forested mountains); Dwight, *Travels in New England and New York*, Vol. 1, 68–74 (71: peculiar); Hale, "Conjectures of the Natural Causes," 61–63 (farther north); Holyoke, "Excels of the Heat and Cold," 65–88; Steele, *The English Atlantic,* 7–9; Chaplin, *The First Scientific American,* 175–178.

CHAPTER 2

1. March 30, 1790, William D. Peck to Jeremy Belknap, Box 1 HUG 1677, WDPP; Lawson, *Passaconaway's Realm*, 54–60; Wilderson, *Governor John Wentworth*, 5, 98; Cutler, "An Account of Some of the Vegetable Productions," Naturally Growing in this Part of America, Botanically Arranged," 395–493, quote on 396; Belknap, *History of New-Hampshire*, Vol 3, 229.
2. Ash, ed., "Expertise," Hindle, *Pursuit of Science*; Stearns, *Science in the British Colonies*; Greene, *American Science*; Parrish, *American Curiosity*.In the mid-twentieth century, American historians looking for the intellectual counterpart to the Industrial Revolution in the North or for the origins of the international prominence of American Cold War science were dismayed by the prevalence of amateurs and the utilitarian focus of eighteenth-century science.
3. Spary, *Utopia's Garden*, quote on 97; McClellan, *Colonialism and Science*; Grove, *Green Imperialism*; Drayton, *Nature's Government*; Harrison, "Science and the British Empire," 56–63; Joseph Banks (JB) to Benjamin Waterhouse (BW), n.d., Benjamin Waterhouse Papers (H MS c16.4), Harvard Medical Library in the Francis A. Countway Library of Medicine [hereafter BW-HMS c16.4]; BW to JB, Dec. 20, 1793, BL Add. Mss. 8098.305–306. For emphasis on the centralizing tendencies of scientific institutions, see Latour, *Science in Action*, Chap. 6.
4. Hindle, *Pursuit of Science*; Stearns, *Science in the British Colonies*; Greene, *American Science*; Chaplin, "Nature and Nation: Natural History in Context," in Prince, ed., *Stuffing Birds*, 76–96; Lewis, "A Democracy of Facts," 663–696; Parrish, *American Curiosity*; Woodward, *Prospero's America*.
5. Exemplary exceptions include Chaplin, *The First Scientific American*; Delbourgo, *A Most Amazing Scene of Wonders*; and Marshall, "Customs in Common," Chap. 14 in *Remaking the British Atlantic*.
6. MacLeod, "Nature and Empire," *Osiris*; "Itineraries of Atlantic Science," *Atlantic Studies*.
7. Harlow, *Founding of the Second British Empire*; Bayly, *Imperial Meridian*; Olson, *Making the Empire Work*; Hancock, *Citizens of the World*; Bowen, *Elites*; Marshall, "Britain Without America—A Second Empire?" in P.J. Marshall, ed., *Oxford History of the British Empire*, Vol. 2, 576–595; Wood, *Radicalism of the American Revolution*, esp. 221–222.
8. Anon., "Letter to the Agricultural Society, dated Nov. 24, 1791," *Nova Scotia Magazine* (February 1792); Cuthbertson, *The Loyalist Governor*, 75; Benjamin Waterhouse to Joseph Banks, December 20, 1793, BL Add. Mss. 8098. 305–306; Williams, *Natural and Civil History of* Vermont,ix-x; Belknap, *History of New-Hampshire*, Vol 3, 229; May 1, 1817, David Humphreys to Joseph Banks, BL Add Ms. 8958. 49–52.
9. Pickering, *A Vocabulary*, 111–112.
10. Mather, *Bonifacius* [title page], 83; March 3, 1766, and Apr. 4, 1766, Cutler and Cutler, eds., *Life, Journals, and Correspondence*, Vol. 1, 12–13; SL47 Harrison Gray

Otis to Charles Lowell, Nov. 10, 1846, Spence-Lowell Collection, Huntington Library; Stowell, *Early American* Almanacs, 72–75; Greene, *American Science*, 63–64; Jaffee, "Village Enlightenment," 327–346; Shields, *Civil Tongues and Polite Letters*; Chaplin, *The First Scientific American*, 41.

11. Eliot, *Essays Upon Field-Husbandry*,26–27; *Rules and Regulations*, 3; Deane, *The New-England Farmer*, 2; Samuel Dexter to John Avery, Aug. 6, 1792, Box 15 Folder, 44, No. 51–64, MSPA-MHS; Thornton, *Cultivating Gentlemen*. On the limited audience for eighteenth-century publications, see Wood, *Revolutionary Characters*, 246–251.
12. Chaplin, *The First Scientific American*; Tyndall, *Lectures on Count Rumford*, 6; Brown, *Benjamin Thompson*, 19, 26, 122–163.
13. Brown, *Benjamin Thompson*; Goode, "The Beginnings of Natural History in America," in Kohlstedt, ed., *Origins of Natural Science*, 23–89; John Adams to Sir John Sinclair, May 24, 1805, in *The Correspondence of the Right Honourable Sir John Sinclair* (London: Henry Colburn and Richard Bentley, 1831), Vol. 2, 38–39. In 1814, the professorship was also subsidized by an annual grant from the state legislature. In 1833, Harvard alumnus and MSPA member Fisher Ames endowed the Fisher Professorship in Botany and Natural History, which replaced the older position On the MSPA's quarrels with Waterhouse and preference for Peck, see Cash, *Dr. Benjamin Waterhouse*, 276–277, 281–283.
14. Draft of inaugural address to Harvard Corporation [1805], Box 2 HUG 1677; March 30, 1790, WDP to Jeremy Belknap, WDPP. On Peck's relationship to the community of agricultural improvers in the Boston area, see Thornton, *Cultivating Gentlemen*, 62–64.
15. Cutler, "An Account of Some of the Vegetable Productions"; Cutler and Cutler, eds., *Life, Journals, and Correspondence*, Vol. 1, 68–69. On Banks, see Gascoigne, *Science in the Service of Empire*; and O'Brian, *Joseph Banks*.
16. [Cutler diary entry], June 14, 1785; Aaron Dexter to MC, June 20, 1785; [Cutler diary entry] June 23, 1785; MC to Samuel Vaughan, Jr., Apr. 10, 1789, in Cutler and Cutler, eds., *Life, Journals, and Correspondence*, Vol. 1, 115–116, Vol. 2, 282; Cutler, *An Explanation of the Map*; Greene, *American Science*, 38–39, 45; Liancourt, *Travels through the United States of North America*, Vol. 1, preface; Onuf, "Cutler, Manasseh," *American National Biography Online* (Feb. 2000).
17. Spary, *Utopia's Garden*; McClellan, *Colonialism and Science*; Latour, *Science in Action*, esp. Chap. 6.
18. Bayly, *Imperial Meridian*; Drayton, "Knowledge and Empire," in Marshall, ed., *Oxford History of the British Empire*, Vol 2, 237; Akita, *Gentlemanly Capitalism*; Caroe, *The Royal Institution*, quotes on 1–2.
19. John Adams to Sir John Sinclair, May 24, 1805, in *The Correspondence of the Right Honourable Sir John Sinclair* (London: Henry Colburn and Richard Bentley, 1831), Vol. 2, 38–39; Jan. 18, 1781, MC to the Harvard Corporation, in Cutler and Cutler, *Life, Journals and Correspondence*, Vol. 1, 82–83.

20. *The Nova Scotia Chronicle and Weekly Advertiser* (Halifax: Anthony Henry, 1769); [Read at a meeting, Oct.31, 1793], Box 14, Folder: 41, No. 13-26, MSPA-MHS.
21. *Letters and Papers on Agriculture: Extracted from the Correspondence of a Society Instituted at Halifax for Promoting Agriculture in the Province of Nova Scotia* (Halifax: John Howe, 1789), 11; Wynn, "Exciting a Spirit of Emulation."
22. May 20, 1796, BW to John Lettsom, BW-H MS c16.4 (emphasis in original); Hugh Findlay to Oliver Smith, Dec. 4, 1793, Box 1, Folder: 11, No. 56-69, MSPA-MHS; Justin Ely to Aaron Dexter, Feb. 6, 1818, Box 1, Folder: 11, No. 41-55, MSPA-MHS; Box 11: Folder 4: Foreign Correspondence, MSPA-MHS.
23. Sir John Temple to JB, Oct. 3, 1787, BL Addl. Mss. 8096.527; Jan. 31, 1797, *Nova Scotia Gazette and Weekly Chronicle*; List of Members, RASE-B.XI and Old Board of Agriculture Letter Books, 1794–1809, RASE-B.XIII, Museum of English Rural Life, Reading, England (MERL); William Strickland to Oliver Smith, March 2, 1797, Box 20, Folder 8: No. 22-42 and Box 1: Folders 6, 11, MSPA-MHS; [Sir John Sinclair's calling card, signed by Sinclair and given to Nova Scotia improver "John Young (Agricola)"], MG 100 Vol. 229 #26, NSA; John Adams to BW, March 11, 1812, *Statesman and Friend: Correspondence of John Adams with Benjamin Waterhouse, 1784–1822*, Worthington C. Ford, ed. (Boston: Little, Brown, 1927); John Adams to Aaron Dexter, May 25, 1813, Folder 45, Box 15, MSPA-MHS; *Laws and Regulations of the Massachusetts Society for Promoting Agriculture: With Extracts of Foreign and Domestic Publications* (Boston: Thomas Andrews, 1793), 35–45; McClellan, *Science Reorganized*, 5, Appendixes 1–2.
24. John Lowell to Massachusetts Society for Promoting Agriculture, June 7, 1813, Folder 1, Box 19, MSPA-MHS; GW to SJS, July 20, 1794, in *The Correspondence of the Right Hon. Sir John Sinclair*, (London: Henry Colburn and Richard Bentley, 1831), Vol. 2, 17–18; Thornton, *Cultivating Gentlemen*, 36–37, 213–215.
25. Timothy Pickering to Oliver Smith, Feb. 12, 1794, Folder 45, Box 15, MSPA-MHS; Folder 10: Miscellaneous, Box 11, MSPA-MHS.
26. Goode, "The Beginnings of Natural History in America," in Kohlstedt, ed., *Origins of Natural Science*, 82; GW to JS, July 20, 1794, *Correspondence of Sinclair*, Vol. 2: 17–18; Cutler to Stokes, Nov. 15, 1793 and May 15, 1805; Cutler and Cutler, eds.,*Life, Journals and Correspondence*, Vol. 2, 275–280.
27. On the history of the Palatine Meteorological Society (1780–1792), see Cassidy, "Meteorology in Mannheim," 8–25.
28. Samuel Williams to JB, Sep. 16, 1789, Dec. 10, 1782–Apr. 27, 1783, SW to Bridgen, Apr. 27, 1783, Box 2, Folder 28, Samuel Williams Papers, Special Collections, University of Vermont (SWP-UV); SW to Solomon Drowne, Jan.31, 1787, Series 1, Box 16, Correspondence, Drowne Family Papers, John Hay Library, Brown University (DFP); Samuel Williams to JB, Sep. 16, 1789, BL Add. Mss. 8097.358; SW to David Rittenhouse [n.d.] and SW to Rumford, Oct. 10, 1786, Series 1: Correspondence, Samuel Williams Family Papers, Ms AM 2624, Houghton Library, Harvard University; McCorison, ed., "A Daybook," 2 (1967), 295 (Stiles);

BW to Joseph Banks (JB), Aug. 10, 1787, British Library Add. Mss. 8096.550, (BL Add. Mss.).

29. Belknap, *A Discourse,* 32; BW to JB, Dec. 20, 1793, BL Add. Mss. 8098.305–306; *Correspondence of Sinclair,* Vol. 2, 6–45; Ellis, *His Excellency: George Washington,* 150; BW to Jean Luzac, Oct. 20, 1805, Benjamin Waterhouse Papers (B MS c10.1), Harvard Medical Library in the Francis A. Countway Library of Medicine. Belknap referred specifically to Cook's voyages, which were a general topic of interest; see also Ezra Stiles to SW, Jan. 19, 1783, SWP-UV.
30. Cash, *Dr. Benjamin Waterhouse*; Fleming, *Science and Technology*; DeLacy, "Fothergill, John (1712–1780)," in *Oxford Dictionary of National Biography Online* (New York: Oxford University Press, 2004).
31. Story, *The Forging of an Aristocracy*; John Adams to BW, Sep. 15, 1812, *Statesman and Friend: Correspondence of John Adams with Benjamin Waterhouse, 1784–1822,* Worthington C. Ford, ed. (Boston: Little, Brown, 1927); John Adams to BW, Aug.7, 1805, *Statesman and Friend,* PP; Nathan Perl-Rosenberg, "Private Letters and Public Diplomacy," 283–311. The lectures were first published as *Heads of a Course of Lectures on Natural History, Now Delivering in the University at Cambridge* (Providence: Bennett Wheeler, 1784–1791), reprinted in Cambridge by Hilliard & Metcalf in 1810; another edition was published as *The Botanist: Being the Botanical Part of a Course of Lectures on Natural History* (Boston: Joseph T. Buckingham, 1811).
32. James Madison to BW, Dec. 27, 1822; Thomas Jefferson to BW, Dec. 1, 1808, March 9, 1813, and July 20, 1816, in Benjamin Waterhouse Papers (BW-H MS c16.2), Harvard Medical Library in the Francis A. Countway Library of Medicine (BW-HMS c16.2); BW to JB, June 15, 1792, BL Add. Mss. 8096.203–204; *Statesman and Friend,* PP; BW to JB, Aug. 10, 1787, BL Add. Mss. 8096.550; Hindle, *Pursuit of Science,* 312; Cash, *Dr. Benjamin Waterhouse,* 89.
33. WPD to Aaron Dexter, London Aug. 10, 1805, WDPP; BW to JB, Aug. 10, 1787, BL Add. Mss. 8096.550; BW to Dudley A. Tyng, Aug. 28, 1806, Box 13, Folder 30, MSPA-MHS. On English sponsors of early American science, see Greene, *American Science,* 9; Stearns, *Science in the British Colonies,* 517–519.
34. Aug. 28, 1790, *Columbian Centinel.* Historians of Maritime Canada typically assume that the War of 1812 ruptured cross-border family and social ties. For one example of continuities, see Conrad, Laidlaw, and Smith, eds. *No Place Like Home,* esp. the diary of Louisa Collins (1797–1869), 61–78; John Halliburton to BW, Oct. 26, 1790, BW-HMS c16.2. Tamara Thornton argues that the flight of many leading Boston families in the Loyalist emigration forced those who stayed behind to re-create their elite identity in Republican terms. Thornton, *Cultivating Gentlemen,* 15–18.
35. "Report of J.B. Palmer on conditions in PEI, 1805," in Series 5, Vol. 15: Lt. Governor PEI, 1804-1812, JDF; miscellaneous correspondence, Titus Smith Papers, MG1 Vol. 1773, NSA; Thomas Brewer to Samuel Pomeroy, March 25, 1807,

Box 20, Folder 7, MSPA-MHS; Gorham, "Titus Smith," 116–123; Terrance M. Punch, "Titus Smith," *Dictionary of Canadian Biography Online.*

36. Dec. 22, 1789, *The Royal Gazette and the Nova-Scotia Advertiser* (Halifax, 1789); William Almon to Oliver Smith, Dec. 12, 1793, Folder: 11, No. 1-19, Box 1, William Almon to Oliver Smith, Dec. 20, 1794, Box 21, Folder 13, MSPA-MHS; Mather Byles to Rebecca Byles Almon, Sep. 23, 1801, MG1, 11: Almon Family Papers, NSA; May 27, 1800, Mather Byles, Jr. (III) to Edward Winslow in W.O. Raymond, ed., *Winslow Papers, 1776–1826* (St. John: New Brunswick Historical Society, 1901), Vol. 2, 447–448; Gilroy, *Loyalists and Land Settlement*, 43. On the close relationship between Mather Byles and Belknap, see Lawson, *The American Plutarch.*
37. Bénigne Charles Fevret de Saint-Mesmin, June 12, 1793 [in *Public Archives of Nova Scotia Report for the Year 1946*, p. *xxvii*], MG1 Vol. 1520 #FF, NSA; Judith Fingard, "Sir John Wentworth," in *Dictionary of Canadian Biography Online.*
38. Rumford to Timothy Walker, May 11, 1775, and Aug. 14, 1776, Folder 7, GB 0116, Benjamin Thompson, Count von Rumford Collection, Royal Institution Archives; Cuthbertson, *The Loyalist Governor*, 7–8, 11, 75; Brown, *Benjamin Thompson*, 96–97, 173–176; Dudley Tyng, "Extract from the Minutes," Oct.30, 1806, Box 2: Papers, 1805-07, WDPP.
39. Sir James Kempt to Charles Ramage Prescott, May 12, 1828, Thomas Cochran Hammill to Maria Prescott, Aug. 25, 1835, Oct. 20, 1818, Prescott Family Papers, Nova Scotia Museum; Thomas Brewer to Samuel Pomeroy, March 25, 1807, Box 20, Folder 7, MSPA-MHS; Samuel Latham Mitchell to JB, Sep. 5, 1795, BL Addl. Mss. 8098.462–463; Peck to [his sister], Sep. 1, 1805, Box 2: Papers, 1805-07, WDP Papers.
40. Lord Dalhousie to Charles R. Prescott, Oct. 20, 1818, Prescott Family Papers, Nova Scotia Museum; Cutler and Cutler, eds., *Life, Journals, and Correspondence*, Vol. 1, 86; Dudley Tyng to WDP, September 30, 1806, WDP Papers; "Report of the committee on ye Garden," July 30, 1808, Box 13, MSPA-MHS; June 1808, WDP to Dudley Tyng, Box 13, Folder 30, MSPA-MHS; Thornton, *Cultivating Gentlemen*, 63–64; Greene, *American Science*, 84.
41. William Aiton to Joseph Banks [Oct. 4, 1803], BL Addl. Mss. 33982.264-265; Henry Bowyer to Joseph Banks, Dec. 23, 1803, BC 2.282, Royal Botanical Gardens, Kew, Library and Archives; [Obituary for Michael Dalton], Feb.12, 1814, *Acadian Recorder*; [Thomas Haliburton (a.k.a. Sam Slick)]; Neale, *The Life of Field Marshall*, 75–80.
42. John Lowell to Mr. Bigelow, Feb. 9, 1809, Box 11, Folder 7: Correspondence, MSPA-MHS (emphasis in original); "Resolve granting a Township of Land to the Massachusetts Agricultural Society," March 4, 1809, Box 13, Folder 30, MSPA-MHS; Cash, *Dr. Benjamin Waterhouse*, 89, 274–279; May 24, 1805, Wastebooks, Vol. 17, Peter Chardon Brooks Papers, Ms N-2049, MHS; Nov. 19, 1807, D.A. Tyng to WDP, WDPP; John Lowell, Committee report, Sep. 24, 1808, Box 13, Folder 30, MSPA-MHS; "Plan for a Professorship of Botany," Feb. 1805, Box 13, Folder 30, MSPA-MHS.

43. "Plan for a Professorship of Botany" and "Vote of the President and Fellows of Harvard College, Nov. 15, 1808," Box 13, Folder 30, MSPA-MHS.
44. "Report of the committee on ye Garden, July 30, 1808"; "Report of the committee on the Garden, July 29, 1809"; "Report of the committee on the Garden, March 31, 1810"; "Report of the committee on the Garden, June 30, 1810"; "Report of the committee on the Garden, July 27, 1811," Box 13, Folder 30, MSPA-MHS; Peck, *Catalogue of American and Foreign Plants,* quote on iii.
45. R.H. Gardiner to Benjamin Vaughan [hereafter BV], March 10, 1805, Folder: Correspondence, January–July 1805, Carton 2: SH 116DN, 1805-1812, Vaughan Family Papers, Correspondence, 1773-1812, MHS (VFP-MHS); "Report on the state of the fund for the professorship, J. Lowell, 1806, Box 13, Folder 30, MSPA-MHS; WDP to Rev. Mr. Kirby, Apr. 22, 1817, Box 2, Folder: Papers, 1815-19, WDPP; Quincy, *History of Harvard University,*Vol. 2, 330, 385–386; WDP to Rev. Mr. Kirby, Apr. 22, 1817, Box 2, Folder: Papers, 1815-19, WDPP; Samuel Drowne to WDP, Aug. 26, 1822, DFP; Thornton, *Cultivating Gentlemen*, 90–91.
46. Marvin, *Benjamin Vaughan,* 1–10; Thorne, *History of Parliament*, Vol. 5, "Members Q–Y"; William Allen, Jr., quoted in Taylor, *Liberty Men and Great Proprietors*, 66.
47. Sarah M. Vaughan (SMV) to Charles Vaughan, Sep.1797; BV to William Manning, Aug. 20, 1797 (emphasis in the original), Carton 1: 1774–1804, Correspondence, 1773–1812, VFP-MHS; Murray, *Benjamin Vaughan.*
48. Aaron Dexter to William Russell, Middletown, Sep.30, 1797, Box 15 Folder: 42, No. 59-72, Massachusetts Society for Promoting Agriculture- MHS; Jeyes, *The Russells of Birmingham*, 195–203; Jacob M. Price, "Russell, William (1740–1818)," *Oxford Dictionary of National Biography Online.*
49. Aaron Dexter to William Russell, Middletown, Sep. 30, 1797, Box 15 Folder: 42, No. 59-72, MSPA-MHS; William Russell to BV, May 3, 1800, Carton 1: Folder: Correspondence, February–May 1800, VFP-MHS; William Russell to BV, May 3, 1800, Carton 1: Folder: Correspondence, February–May 1800, VFP-MHS; BV to Samuel & Sarah Vaughan, Sep. 25, 1798, Carton 1: Folder: Correspondence, July–September, 1798, VFP-MHS; BV to Samuel and Sarah Vaughan, Sep. 25, 1798, VFP-MHS; Murray, *Benjamin Vaughan*, 384–387; Graham, "Joseph Priestley in America," in Rivers and Wykes, eds.,*Joseph Priestley*, 203–230.
50. SMV to Charles Vaughan, Sep. 1797, Carton 1: 1774-1804, VFP-MHS; BV to William Manning, Aug. 20, 1797, Carton 1: 1774–1804, VFP-MHS; SMV to Charles Vaughan, Sep. 1797; Henry Bird to SMV Apr. 4, 1796; Eliza Bird to SMV, December/January 1796/7, Folder: Correspondence, April–May 1796; BV to Samuel and Sarah Vaughan, Sep. 27, 1797, Carton 1: 1774–1804, VFP-MHS.
51. Augusta, Princess of Wale's Accounts, Box 5: Nos. 55422-55661, Georgian Archive, Royal Archives, Windsor; Phillips, *The Scientific Lady*; Valenze, "The Art of Women," 142–169. By the nineteenth century, the amateur pursuit of botany and horticulture became increasingly feminized. See Shteir, *Cultivating Women, Cultivating Science.*

52. Eliot, *Essays Upon Field-Husbandry*, 9; Deane, *The New England Farmer*, 47–49; Shields, *Civil Tongues and Polite Letters*, 104–119.
53. Cutler to Mrs. Torrey, Nov. 22, 1803 and Feb. 21, 1805; Merry to Cutler, May 17, 1806, in Cutler and Cutler, eds., *Life, Journals and Correspondence*, Vol. 2, 144–146, 190; [Ann?] Blakey Sharpless [n.d./early 19th century], in *New England Women and the Families in the 18th and 19th Centuries*, Series B: Manuscript Collections from the Newport Historical Society, 10: 0314; Parrish, *American Curiosity*, 136–173.
54. RR to TS, Nov.30, 1800, RR to TS, Jan. 23 1802, miscellaneous correspondence, Titus Smith Papers, MG1 Vol. 1773, NSA. On British improving families in pre-Confederation Quebec, see Coates, *Metamorphoses of Landscape and Community*, 149–152.
55. Cutler to Jeremy Belknap, March 19, 1789, Cutler and Cutler, eds., *Life, Journals and Correspondence*, Vol. 2, 255; Henry Bird to SV, Apr. 4, 1796; Eliza Bird to SV, Dec. 1796 and Jan. 1797; John Vaughan to BV & SV, Aug.17, 1799; Samuel Vaughan to BV, Feb. 28, 1798; and [BV] to "Samuel" [West Indies], June 1, 1803, Carton 1, VFP-MHS. Benjamin compared owning property in Jamaica to "planting a vineyard . . . at the foot of an active volcano; . . . it is like a man of ninety marrying a giddy young girl practised in the arts of poisoning."
56. William O. Vaughan to Henry Vaughan, Henry Vaughan to BV, c. 1801; [Draft] William O. Vaughan to John B. Bordley, Oct. 1801; Bordley to W. Vaughan, Nov. 1, 1801, Carton 1, VFP-MHS.
57. J.P. Anderson to BV, July 27, 1798, Carton 1; [sibling] to SV, ca. 1798 (emphasis in original); SMV to CV, Sep. 1797, Carton 1, VFP-MHS; William Manning, Totteridge, England, to SMV, Feb. 1798, Carton 1, VFP-MHS.
58. Oct. 15, 1798, E.M. Bird to Benjamin Vaughan, Carton 1, VFP-MHS.
59. Ulrich, *A Midwife's Tale*, 29, 96, 311–329, 384, fn. 37; BV's postscript in SMV to CV, Sep. 1797, Carton 1, VFP-MHS, emphasis in original.
60. *The Nova Scotia Chronicle and Weekly Advertiser* (Halifax: Anthony Henry, 1769); William Shaw to Colonel Edward Winslow, Jan. 1, 1785, C.B. Ferguson Papers, MG 1 Box 1898, Folder 9, NSA; Mason, "The American Loyalist Diaspora and the Reconfiguration of the British Atlantic World," in Gould and Onuf, eds., *Empire and Nation*, 239–259; Calhoon, Barnes, and Rawlyk, eds., *Loyalists and Community*, 6–9.
61. Faragher, *A Great and Noble Scheme*, 443–454; Hodson, *The Acadian Diaspora*, 70–78, 198.
62. Council for Nova Scotia, Report to the King, Sep. 27, 1720, Folder O, MG1, Vol. 1520 [copy of PRO CO 217/4, WA 53], NSA; "Information respecting the Estate of Tatmagouche in Nova Scotia obtained on the spot by Capt. John Macdonald and transmitted to his friend L. Gov'r Desbarres the Proprietor thereof." Series 2 (MG 23, F1-2), Joseph Frederick W. DesBarres Papers, 1762–1894, MG23, F1, National Archives of Canada; Titus Smith, "Survey of the Eastern and Northern Parts

of the Province in the Years 1801 & 1802 . . . and Observations on the Western Parts," RG1 380A, NSA; Choquette, *Frenchmen into Peasants*, 277–290.

63. Aug. 15, 1805, William Vaughan to Dr. Joseph McKeon, Carton 2, VFP-MHS.

CHAPTER 3

1. "Canada Survey'd" (unpublished manuscript, 1708), Samuel Vetch papers, 1708–1712, Rare Book and Manuscript Library, Columbia University; "Canada Survey'd, or the French Dominions upon the Continent of America briefly considered," July 27,1708, CO 323/6, Nos. 64, 65 and CO 324/9; Alsop, "Samuel Vetch's 'Canada Survey'd,'" 39–58, and "The Age of the Projectors," 30–53. Alsop (1982) usefully discusses the differences between successive drafts of "Canada Survey'd."
2. Governor Dudley to the Council of Trade and Plantations, Oct. 2, 1706, CO 5/864, Nos. 114-126 (Scotch climate); McNeill, *Mosquito Empire*, 91–136 (mortality figures on 118–119).
3. William Dandridge Peck, "Lectures on Botany," HUG 1677.53, WDPP (quotation). For acclimatization, see Chaplin, *Subject Matter*, Chap. 4; Anderson, "Climates of Opinion," 135–157; Spary, *Utopia's Garden*, 117–125; Koerner, *Linnaeus*; Osborne, "Acclimatizing the World," 135–151; Ritvo, "Going Forth and Multiplying," 404–414; Jonsson, *Enlightenment's Frontier*, 81–88; Bashford, *Global Population*, 147–153, 202–206.
4. Davenant, *Ways and Means of Supplying the War*, 140–141, 149; idem, *Some Objections Against Settling the Trade to Africa*, 2; idem, *Constitution and Management of the Trade to Africa*, 37; idem, *Universal Monarchy*, 58–59; Ted McCormick, "Population: Modes of Seventeenth-Century Demographic Thought," in Stern and Wennerlind, *Mercantilism Reimagined*, 25–45; Statt, *Foreigners and Englishmen*, Chaps. 3 and 4. For the argument that climate was crucial to understanding a series of crises in the seventeenth-century Ottoman Empire, see White, *Climate of Rebellion*, 123–225.
5. Petty, *Political Arithmetick*, 90, 94, 96–97; Newell, *From Dependency to Independence*, 3; McCormick, *William Petty*, 235–236; Graham, *British Policy and Canada*.
6. This comparison, published in 1780, essentially excerpted and put back into circulation Chaps. 25 and 27 from William Burke's *An Account of the European Settlements in America*, first published in 1757, which had also included the New England colonies in its condemnation of the northern climate. *A New Collection of Voyages, Discoveries and Travels, containing whatever is worthy of notice* (London, 1767), Vol. 2, 133; *A True Account of the Colonies of Nova Scotia and Georgia* (Newcastle, Printed in the present year [1780?]), 7, 21; McElroy, "Burke, William (1728–1798)," in *Oxford Dictionary of National Biography Online* (New York: Oxford University Press, 2004). Elizabeth Lichtenstein Johnston

(1764–1848), [unpublished autobiography, n.d.], Almon Family Papers, Series B, NSA; After living in South Carolina, New York, Jamaica, and Scotland, in 1806 she and her husband eventually joined her father in what she described as the "cooler climate" of Nova Scotia.

7. Young, *Present State of the Wastelands of Great Britain*, 10–13; idem, *Annals of Agriculture*; Betham-Edwards, ed., *The Autobiography of Arthur Young*, 61, 83 (mother, emigration).
8. John Wentworth to Earl of Balcarres, Lieutenant Governor of Jamaica, 22 July 1798, RG 1 Vol. 52/7 1796-1800, Wentworth Papers, NSA.
9. John Smith, *A Description of New England*, 16–17; Best, *Three Voyages of Martin Frobisher*, 4–6; Kupperman, "Fear of Hot Climates," 216 (Gorges); Morton, *New English Canaan*, 83.
10. Gosnold quoted in Quinn and Quinn, eds., *English New England Voyages*, 209; Brereton, *Discouerie of the North Part of Virginia*, 11; [William Bradford], *A Relation or Iournall of the Beginning and Proceedings of the English Plantation Setled at Plimoth in New England* (London, 1622), 13–14.
11. Hubbard, *General History of New England*, 29; Cronon, *Changes in the Land*; Lanphear and Snow, "European Contact and Indian Depopulation," 15–33; Salisbury, *Manitou and Providence*; Crosby, *Ecological Imperialism*; Cronon, *Changes in the Land*; Anderson, *Creatures of Empire*; Chaplin, *Subject Matter*, Chaps. 1–4; Jones, *Rationalizing Epidemics*.
12. Wood, *New England's Prospect*, 15; May 22, 1634, John Winthrop to Nathaniel Rich, Gilder Lehrman Collection #: GLC01105; https://www.gilderlehrman.org/sites/default/files/inline-pdfs/ready.FPS_01105.pdf; Vincent, *A True Relation of the Late Battell*, 20; Chaplin, "Natural Philosophy and an Early Racial Idiom," 229–252; idem, *Subject Matter*, 151–152.
13. Conforti, *Imagining New England*, Higginson quoted on 25; June 16, 1795, *The Royal Gazette and the Nova-Scotia Advertiser*; Grove, *Green Imperialism*, 156–161; Belknap, *History of New Hampshire*, Vol. 3, 17.
14. Morton, *New English Canaan*, 15–16; Vincent, *A True Relation of the Late Battell*, 21–22; Denys, *Description géographique et historique*, n.p. ("Le climat est pareil au nostre et sous la mesme elevation. Il est plus facile a peupler qu'aucune des terres de l'Amerique ou nous avons des colonies, parce que le voiage en est court, et se fait presque tout enteir sous le mesme parallele d'ou l'on a coustume de partir pour y aller.")
15. Daniel Smith, "Demographic History of New England," 165–183; Reid, *Acadia, Maine, and New Scotland*, 20–25; idem, *Essays on Northeastern North America*, 23–68; Greene, *Pursuits of Happiness*, 178–179; Rawlyk, *Nova Scotia's Massachusetts*, preface, 208–209; Lockridge, "Land," 62–80; Bushman, *From Puritan to Yankee*, 257–259; Kulikoff, *From British Peasants to Colonial American Farmers*, 113–115, 153–154, 239–240; Jaffee, *People of the Wachusett*, 3–7. For a nuanced view of soil exhaustion, see Donahue, *The Great Meadow*.
16. Moore, *Pilgrims*, 1–14, 76, 102; James Horn and Philip D. Morgan, "Settlers and Slaves: European and African Migrations to Early Modern British America,"

Mancke and Shammas, eds., in *Creation of the British Atlantic World*, 23, 25; Games, *Migration*; Choquette, *Frenchmen into Peasants*; McCusker and Menard, *The Economy of British America,* 101–107, 216–227; Bailyn, *Voyagers to the West*, 371–372, 460–461; idem, *New England Merchants*, 1–2.

17. Feb. 10, 1786, Cutler to Belknap; March 1, 1786, Belknap to Cutler; Cutler and Cutler, eds.,*Life, Journals and Correspondence*, Vol. 2, 239–240; Cayton, *The Frontier Republic*, Chap. 1.
18. *A True Account of the Colonies of Nova Scotia and Georgia* (Newcastle: Printed in the present year [1780?]), 7; Bumsted, *The People's Clearance*, 15 (fields); Anon., *American Husbandry: Containing an Account of the Soil, Climate, Production, and Agriculture* (London: J. Bew, 1775), 1; James Horn, "British Diaspora: Emigration from Britain, 1680-1815," in Marshall, ed., *Oxford History of the British Empire*, Vol. 2, 29–32; O'Grady, *Exiles and Islanders*. Nova Scotia absorbed the most Scottish emigrants after 1783, predominantly Highlanders; see Horn, "British Diaspora," 32.
19. Bailyn, *The New England Merchants*; Vickers, *Farmers and Fishermen*, 271–275; McCusker and Menard, *The Economy of British America,* 113–114; Reid, *Acadia, Maine, and New Scotland*; Plank, *An Unsettled Conquest*, Chap.2; Johnson, *John Nelson,* Chap. 4.
20. *Caledonia Triumphans: A Panegyrick to the King* (Edinburgh, 1699) (pinching); Borland, *Memoirs of Darien*, 100 (a Place too hot).
21. Plank, *An Unsettled Conquest*, 49–58; Haefeli and Sweeney, *Captors and Captives*, Chap. 9.
22. "Canada Survey'd" (unpublished manuscript, 1708), Samuel Vetch papers, 1708–1712, Rare Book and Manuscript Library, Columbia University; for combined forces see, "An Explanatory Supplement to Capt. Vetch's Proposal for an Attack upon Quebec and Montreal," Nov. 17, 1708, CO 323/6, No. 71 and CO 324/9; Alsop, "Samuel Vetch's 'Canada Survey'd,'" 39–58; idem, "The Age of the Projectors," 30–53.
23. Samuel Vetch, "A Poem," 5–7; John G. Reid, "1668-1720: Imperial Intrusions," in Buckner and Reid, eds., *Atlantic Region to Confederation*, 90–91.
24. "Nicholson and the Council of War at Annapolis Royal to the Queen," [?] Oct. 1710, CO 5:9; "Vetch, Governor of Annapolis, to the Council of Trade and Plantations," 26 Nov. 1711, CO 217:1, No. 1 and CO 218:1; Statt, *Foreigners and Englishmen*, Chaps. 5 and 6; Otterness, *Becoming German*; Leder, *Robert Livingston,* 211–246; Knittle, *Palatine Emigration*, 156–170; Dobson, *Scottish Emigration*, 85–86.
25. "Col. Vetch to the Council of Trade and Plantations," Dec. 22, 1714, CO 217:1, no. 21; Samuel Vetch to the Lords of Trade, Feb. 24, 1715, MG1, Vol. 1520, Folder L, NSA.
26. "H.M.'s Warrant," June 23, 1713, CO 217:1, No. 19 and CO 218:1; Thomas Caulfeild to Board of Trade, Nov. 21, 1715, MG1, Vol. 1520, Folder K, NSA [copy of PRO CO 217:2, WA 53]; "Col. Vetch to the Council of Trade and Plantations," Feb. 21, 1716, CO 217:2, no. 16 and CO 218:1. For a detailed examination of the bitter internal

disputes at the garrison, including those between Vetch and Nicholson, see Plank, *An Unsettled Conquest*, 54–57.

27. Hodson, *The Acadian Diaspora*, Chap. 1; Griffiths, *From Migrant to Acadian*, 285; Bailyn, *Voyagers to the West*, 362–364; O'Grady, *Exiles and Islanders*; Bell, *The 'Foreign Protestants,'* 18–28; *Censuses of Canada, 1665 to 1871, Statistics of Canada*, Vol. 4, accessed online: http//www.statcan.gc.ca/pub/98-187-x/4064810-eng.htm (13,000; 9,000); Stephen White, "The True Number of the Acadians," in Ronnie-Gilles LeBlanc, ed., *Du grand dérangement à la déportation*, 21–56.
28. *Scots Magazine* 34 (Sep. 1772), 482–484; *Scots Magazine* 34 (Nov. 1772), 587–588; Bumsted, *The People's Clearance*, 56–73; Bailyn, *Voyagers to the West*, 390–397.
29. Deschamps, "Observations on the Progress of Agriculture in Nova Scotia and New Brunswick with Notices of Acadian Manners and Customs . . ." (1782), RG 1, Vol. 284, NSA; [DesBarres's draft history of the controversy in PEI], Vol. 14, Series 5: Additional correspondence and papers, 1762–1894, J.F.W. DesBarres Papers, 1762–1894, MG23-F5, National Archives of Canada [NAC] (not settlers); "The History of the Province of Nova Scotia," (unpublished manuscript, 1790), Papers of Professor Andrew Brown (1763–1834), Edinburgh University Library, Special Collections (one hundred). Formulated after the Quebec Act of 1774, Deschamps's proposal may have also reflected new ideas about sectarian settlement policies in formerly French areas of British North America.
30. George Deschamps and G.H. Monk proposal for purchase of Lord William Campbell's land in Windsor, May 17, 1783, J.F.W. DesBarres Papers, Series 5: Additional correspondence and papers, 1762–1894, MG 23, F1-5, Mary Cannon, NAC; Jasanoff, *Liberty's Exiles*, 352–353 (estimates of Loyalist influx).
31. Cape Breton Governor's Accounts, 1784–1801, J.F.W. DesBarres Fonds, Series 3, 1774–1807, MG 23, F1-3, National Archives of Canada [hereafter JDF]; "Letter of Edward Winslow to the *Royal Gazette*, July 1802," in W.O. Raymond, ed. *Winslow Papers, 1776–1826* (St. John: New Brunswick Historical Society, 1901), Vol. 2, 469. On DesBarres, see Evans, *Uncommon Obdurate*; and Hornsby, *Surveyors of Empire*.
32. [DesBarres's draft history of the controversy in PEI], Vol. 14, Series 5: Additional correspondence and papers, 1762–1894, JDF; DesBarres quoted in "The History of the Province of Nova Scotia," (unpublished manuscript, 1790), Papers of Professor Andrew Brown (1763–1834), Edinburgh University Library, Special Collections; Oct. 22, 1791, Rory Steel to Colin MacDonald, Royal Highland & Agricultural Society of Scotland Papers, Adv. MS.73.2.13, f.27, National Library of Scotland.
33. "Report of J.B. Palmer on conditions in PEI, 1805," in Series 5, Vol. 15: Lt. Governor PEI, 1804-1812, JDF; Peck, "Introductory Botanical Lecture," WDP, Box 2: Folder 1805-07; Benjamin Vaughan to Judge Lowell, March 19, 1800, Carton 22: SH 116Y1, Folder: Misc. Agricultural Papers, Vaughan Family Papers, MHS; Shipton, *Biographical Sketches*, Vol. 13, 19 (Byles, Siberia). Thanks to Edward M. Griffin for the Byles citation.

34. Kalm, *Travels into North America*, Vol. 1, 106; William Peck to Lydia Peck, April 1806, Box 2: Folder 1805-07, WDPP; Peck to Lydia Peck, Apr. 1806, WDP, Box 2: Folder 1805-07; Dwight, *Travels in New England and New York*, Vol. 1, 19–21, Vol. 2, 213; Jonsson, *Enlightenment's Frontier*, 5–6.
35. Deane, *The New-England Farmer*, 45, 58, 108; Dickson, *A Treatise of Agriculture*; Belknap, *The History of New-Hampshire*, Vol. 3, 120.
36. Robinson and Rispin, *Journey Through Nova-Scotia*, 22; [Fragment of a letter to James Clarke, Secretary of NSSPA], [n.d.], FOLDER 5, item 27, MG1 Vol. 1664A-B, NSA; *Letters and Papers on Agriculture in the Province of Nova Scotia*, 9.
37. Paul Mascarene, "Description of Nova Scotia," 1720/21, MG1, Vol. 1520, Folder P, Mascarene (Engineer), NSA; [Otis Little], *The State of Trade in the Northern Colonies Considered, with an Account of their Produce and a Particular Description of Nova Scotia* (London, 1748), 34–35; Sep. 27, 1797, Benjamin Vaughan to Samuel and Sarah Vaughan, Carton 1, VFP-MHS; Asher Robbins, *An Address to the Society for the Promotion of Agriculture and other Useful Arts in the state of Rhode-Island and Providence Plantations* (Newport, RI: 1802); April 8, 1815, David Waldron to MSPA, MSPA Box 15: Folder: 44, No. 41–50, MHS; *Letters and Papers on Agriculture in the Province of Nova Scotia*, 9. On ecological distinctions between northern and southern New England, see Cronon, *Changes in the Land*, 44, 50, 86–87.
38. A Farmer to Mr. Clark, Secretary of the Agricultural Society, [n.d.], Folder 5, item 26, MG1 Vol. 1664A-B, NSA; *Letters and Papers on Agriculture . . . in the Province of Nova Scotia*, 9; Ruggles to Edward Winslow, July 17, 1783, MG 100, Vol. 216, No. 15-17, NSA; Ruggles, *General Timothy Ruggles*, 20–21.
39. Belknap, "On the Preserving of Parsnips," 199–200; Strickland, *Journal of a Tour*,204; Jeyes, *The Russells of Birmingham*, 231–232. See Pauly, "Fighting the Hessian Fly," 377–400; and Brooke Hunter, "Creative Destruction: The Forgotten Legacy of the Hessian Fly," in Cathy Matson, ed., *The Economy of Early America*, 236–262.
40. Belknap, *History of New-Hampshire*, Vol. 3, 17; Hawley, "The Agency of God," 7–8; "Notes on the State of Vermont," Samuel Williams Family Papers, Ms. AM 2624, Houghton Library.
41. "Oxen," [n.d.] Benjamin Waterhouse Papers (H MS c16.4), Harvard Medical Library in the Francis A. Countway Library of Medicine (BW-HMS c16.4); Rothenberg, "The Emergence of a Capital Market," 781 (dismal); Main and Main, "Economic Growth," 27–46; Buckner and Reid, *Atlantic Region to Confederation*, 19–20 (precolonial horticulure); Vickers, "The Northern Colonies," 209–248; Donahue, *The Great Meadow*; McWilliams, "'To Forward Well-Flavored Productions,'" 25–50; idem, *Building the Bay Colony*.
42. [Fragment of a letter to] Mr. James Clarke, Secretary of the Society for Promoting Agriculture, Folder 5, item 27, Titus Smith Papers, MG1 Vol. 1664B, NSA (saffron); Governor the Earl of Bellomont to the Council of Trade and Plantations, Nov. 28, 1700, CO 5/1083, Nos. 49.

43. Eliot, *Essays Upon Field-Husbandry in New England and Other Places,* 127–129, 134–135, 143; Ames, *An Astronomical Diary.* Sericulture schemes also largely failed in the eighteenth-century South, and silk never became a significant American export.
44. J.C. [Lettsom?] to Oliver Smith, November 18, 1793, Box 11: MSPA, MHS (letter marked "not to be published"); "American Produce," Sep. 1, 1790, *Columbian Centinel*; James Winthrop to MSPA, June 26, 1793, Box 2, Folder: 11, MSPA, MHS; Benjamin Waterhouse "Lecture: The Silkworm" Dec. 4, 1810 HMS c16.4; Koerner, *Linnaeus,* 149; Jonsson, *Enlightenment's Frontier,* 60, 83, 99; Chaplin, *Anxious Pursuit,* 158–116. On the repeated attempts to promote hemp and silk as specialty crops in New England and their failure, see Bidwell and Falconer, *History of Agriculture,* 100–101, 193.
45. May 30, 1752, *Halifax Gazette*; Feb. 19, 1805, SR RASE BIII: 1802–1803, MERL; "Plan of a Society for Promoting Agriculture in the Province of Nova Scotia"; "A New Method of Cultivating and Preparing Hemp . . . *Printed by order of the Lords of the Committee of Council of Trade and Foreign Plantations,*" Apr. 1790, *Nova Scotia Magazine*; [Edward Long], *Free and Candid Review of a Tract entitled "Observations on the Commerce of the American States with Europe and the West Indies"* (London, 1784), 39, 45 (transient, shriveled).
46. *Letters and Papers on Agriculture: Extracted from the Correspondence of a Society Instituted at Halifax for Promoting Agriculture in the Province of Nova Scotia,* Vol. 1 (Halifax: John Howe, 1789), 10; Nov. 3, 1789, *The Royal Gazette and the Nova-Scotia Advertiser*; Nov. 1789, *Nova Scotia Magazine*; [John Holroyd Sheffield], *Observations on the Commerce of the American States with Europe and the West Indies* (London, 1783); Drayton, *Nature's Government,* 115–120.
47. "The History of the Province of Nova Scotia," (unpublished manuscript, 1790), Papers of Professor Andrew Brown (1763–1834), Edinburgh University Library, Special Collections; Brown to Belknap, Dec. 31, 1792, Brown to Belknap, Feb. 25, 1797, *Belknap Papers MHSC,* 537, 613; Cape Breton Governor's Accounts, 1784–1801, JDF, Series 3, 1774–1807, MG 23, F1-3 (useful and loyal).

    Brown did not complete or publish this manuscript. After returning to Scotland, he became a professor of rhetoric and literature at the University of Edinburgh. For a biographical sketch of Brown and an analysis of his account of the *grande derangement,* see Beanlands, "Annotated Edition of Rev. Dr. Andrew Brown's Manuscript."

CHAPTER 4

1. John Wentworth (JW) to Reverend Dr. Morris, May 1, 1799, Wentworth Papers, RG 1 Vol. 52/7, 1796–1800, Nova Scotia Archives (WP-NSA).
2. Christopher, "Disgrace to the very Colour"; Fergus, "'Dread of Insurrection,'" 757–780, on the Chinese immigration scheme, see pp. 776–778.

3. Curtin, "'The White Man's Grave,'" 94–110; Harrison, "'The Tender Frame of Man,'" 68–93; Anderson, *The Cultivation of Whiteness*; McNeill, *Mosquito Empires*; Webb, *Humanity's Burden*.
4. Kopytoff, "Early Jamaican Political Development," 287–397; Geggus, "The Enigma of Jamaica," 274–299; Orlando Patterson, "Slavery and Slave Revolts: A Sociohistorical Analysis of the First Maroon War, 1665-1740," in Price, ed., *Maroon Societies*, 246–292.
5. *The Proceedings of the Governor and Assembly of Jamaica, in regard to the Maroon Negroes* (London, 1796); Dallas, *History of the Maroons*; Balcarres to Barham [Middleton, Charles, first Baron?], May 23, 1796, CO 137/97, "Correspondence regarding Jamaica," Vol. 2, May–December 1796; Michael J. Day, "Military Campaigns in Tropical Karst: The Maroon Wars of Jamaica," in Caldwell and Harmon, eds., *Studies in Military Geography*, 79–88.
6. *Proceedings of Jamaica, in regard to the Maroons*, 52.
7. Dallas, *History of the Maroons*, Vol. 2, 181 (disposed), 195-196 (ocean).
8. JW to Morris, May 1, 1799, RG 1 Vol. 52/7, WP-NSA; Dallas, *History of the Maroons*, Vol. 2, 207–209; JW to Duke of Portland, July 25, 1796, *Extracts and Copies of Letters from Sir John Wentworth, Lieutenant Governor of Nova Scotia, to his Grace the Duke of Portland, respecting the Settlement of the Maroons in that Province* (London, 1797), 4–5.
9. JW to Evan Nepcan, Dec 13, 1792, RG 1 Vol. 50/5, 1792–93, WP-NSA; *Extracts and Copies Wentworth to Portland*, 5; Apr. 21, 1797, JW to Duke of Clarence, RG 1 Vol. 52/7, 1796–1800, WP-NSA. The fortified section they completed was known as Maroon Bastion until the citadel built in the mid-nineteenth century replaced it. Naftel, *Prince Edward's Legacy*, 48. By comparison, the average daily wage for common laborers in Nova Scotia during the 1790s was CAD$0.83, roughly 2 shillings—more than twice the Maroons' wages. For a table of average daily wages, see Gwyn, *Excessive Expectations*, 23.
10. *Eighteenth Parliament of Great Britain: First Session* (Sep. 27, 1796–July 20, 1797), Apr. 6–May 1, 1797, Vol. 52; *The Parliamentary Register; or Debates and Proceedings of the House of Commons* (London, 1798), 407 [live].
11. Sep. 25, 1796, JW to Lord Fitzwilliam; Sep. 24, 1796, JW to Portland, WP-NSA.
12. JW to Portland, Aug. 13, 1796; Aug.15, 1796, JW to J. Eginton, RG 1 Vol. 51/6 1793–1796, WP-NSA; Grant, *The Maroons in Nova Scotia*, 7; *Eighteenth Parliament of Great Britain: First Session* (Sep. 27,1796–July 20, 1797), Apr. 10, 1797, Vol. 52, 472 (Wentworth-Portland letters).
13. JW to Portland, Apr.21, 1797; JW to Duke of Clarence, Apr. 21, 1797; JW to Portland, July 22, 1799, RG 1 Vol. 52/7, 1796–1800, WP-NSA; *Eighteenth Parliament of Great Britain: Second Session* (Nov. 2, 1797–June 29, 1798), Feb. 14–15, 1798, 297; 307 (shot); Fyfe, *History of Sierra Leone*, 80.
14. Apr. 21, 1797, JW to Duke of Clarence, RG 1 Vol. 52/7 1796–1800; Frances Wentworth to Lady Fitzwilliam, Oct. 5, 1784 and Dec. 20, 1799, Governor John Wentworth Fonds, MG 1, Nova Scotia Archives, WF-NSA.

15. Wood, *New England's Prospect*, 4; Neal, *History of New England*, Vol. 2, 565 (exercise); Hubbard, *A General History of New England*, 20; John Adams to Benjamin Waterhouse, Feb. 19, 1805, *Statesman and Friend* [add pp]; Jan. 26, 1730, *New-England Weekly Journal*, 2; Dwight, *Travels in New England*, Vol. 1, 86; Anderson, *New England's Generation*, 225 (average ages at death).
16. Apr. 21, 1797, JW to Duke of Clarence, RG 1 Vol. 52/7, 1796–1800, WP-NSA.
17. JW to Portland, Apr. 21, 1797, JW to Duke of Clarence, Apr. 21, 1797, RG 1, Vol. 52/41d, WP-NSA; Jan. 31, 1797, *Royal Gazette and Nova Scotia Advertiser* [seed]; June 27, 1797, *Royal Gazette and Nova Scotia Advertiser* [Quarrell posted a notice for hiring out Maroons].
18. Apr.21, 1797, JW to Portland; Apr. 21, 1797, JW to Duke of Clarence; June 2, 1797, JW to Portland, RG 1 Vol. 52/7, 1796–1800, WP-NSA; Grant, *The Maroons in Nova Scotia*, 60–68, 118–120; Winks, *The Blacks in Canada*, 81 (number at Boydsville). On the Sandemanians (or Glassites) in North America, see Cantor, *Michael Faraday*, 5, 94–95; Smith, *The Perfect Rule of the Christian Religion: A History of Sandemanianism*, 132 (ten families).
19. JW to Quarrell, Sep. 10, 1797; JW to Earl of Balcarres, Dec. 20, 1797, and July 22, 1798; JW to Portland, June 23, 1798, and April [n.d.] 1799, RG 1 Vol. 52/7, 1796–1800, WP-NSA; Dallas, *History of the Maroons*, Vol. 2, 252–256 (Maroon letter),477–505, Grant, *The Maroons in Nova Scotia*, 83–87.
20. July 22, 1798, JW to Earl Balcarres; April [n.d.] 1799, JW to Portland, RG 1 Vol. 52/7, WP-NSA; Winks, *The Blacks in Canada*, 85 [cabbages].
21. Scott, *Domination and the Arts of Resistance*.
22. Brymer, "The Jamaica Maroons—How They Came to Nova Scotia—How They Left It," *Foreign and Commonwealth Office Collection* (1895), 90. For the history of black Loyalists across the Atlantic world, see Walker, *The Black Loyalist*; Schama, *Rough Crossings*; Campbell, ed., *Back to Africa*, 2–32.
23. Sierra Leone Company, *An Account of the Colony of Sierra Leone From its First Establishment in 1793: Being the substance of a report delivered to the proprietors* (London, 1795); House of Commons, "Report from the Committee on the Petition of the Court of Directors of the Sierra Leone Company" (London, 1802), 735–746, quote on 737; JW to John Gray, Aug. 5, 1800, RG 1 Vol. 53/8, 1800–1805, WP-NSA; Jasanoff, *Liberty's Exiles*, 172–175, 279–309.
24. Dallas, *History of the Maroons*, Vol. 2, 199.
25. Clifton, ed., *Hippocrates Upon Air, Water, and Situation*, 33–34, 37; Floyd-Wilson, *English Ethnicity and Race* ("white frosty" and "scorched" skin, quoted on 30); on the persistence of humoral theory of disease into the mid-nineteenth century, see Valencius, *The Health of the Country*, esp. 53–108.
26. Kupperman, "Fear of Hot Climates," 213–240; Chaplin, *Subject Matter.*
27. Greene, "The American Debate," 384–396; Curran, *The Anatomy of Blackness*; Mitchell, "Causes of the Different Colours," 102–150; Delbourgo, "The

Newtonian Slave Body," 185–207; Long, *The History of Jamaica*, Vol. 3, 354–356; Harrison, *Medicine in an Age of Commerce*, 71–73, 108–110.

28. Arbuthnot, *Concerning the Effects of the Air*, 115–128 (on diseases born of cold air); Lind, *Diseases Incidental to Europeans*, 19–29; Black, *Mortality of the Human Species*, 120–122, 393–400; Rogers, *A Concise Account of North America*, 18–19 (northern provinces).
29. Anon., *American Husbandry: Containing an Account of the Soil, Climate, Production, and Agriculture* (London: J. Bew, 1775), 1–3; Dallas, *History of the Maroons*, Vol. 2, 198–199.
30. [untitled, n.d.], Item 28, Folder 5, MG1 Vol. 1664B, TSP-NSA.
31. Smeathman, *Plan of a Settlement.* On Smeathman, Sierra Leone, and the black Loyalists, see Coleman, *Romantic Colonization*, Chaps. 1 and 3; and Douglas, "The Making of Scientific Knowledge."
32. "Queries Respecting the Slavery and Emancipation of Negroes in Massachusetts, Proposed by the Hon. Judge Tucker of Virginia, and Answered by the Rev. Dr. Belknap," *Collections of the Massachusetts Historical Society*, Ser. 1, Vol. 4, (Boston: MHS, 1835 [1795]), 199.
33. Pennant, *Arctic Zoology*, cclxxxvi. Benjamin Thompson (Count Rumford) offered differences in skin color in hot versus cold climates as an illustration of his theory of heat; see Brown, *Collected Works of Count Rumford*, Vol. 1, 319–322.
34. Smeathman, *Plan of a Settlement*, 10–11.
35. House of Commons, "Report . . . of the Sierra Leone Company," 744; Campbell, *Back to Africa*, 23–24 (remembrance).
36. *Eighteenth Parliament of Great Britain: Second Session* (Nov. 2, 1797–June 29, 1798), May 1, 1798, 90–93 (Liguanga; gentlemen); *Papers Relative to the Settling of the Maroons in His Majesty's Province of Nova Scotia* (London, 1798), 11–13 (less tolerable).
37. Philip D. Morgan and Andrew J. O'Shaughnessy, "Arming Slaves in the American Revolution," in Brown and Morgan, *Arming Slaves*, 191.
38. JW to Portland, Aug. 13, 1796; Aug. 16, 1796, JW to George Hammond, RG 1 Vol. 51/6, 1793–1796, WP-NSA.
39. JW to Lord Fitzwilliam, Sep. 25, 1796, Governor John Wentworth Fonds, MG 1, Nova Scotia Archives, WF-NSA.
40. JW to Portland, Oct. 22, 1799, RG 1 Vol. 52/7; JW to Portland, Oct. 29, 1796, RG 1 Vol. 51/6, 1793–1796, WP-NSA.
41. Greene, *Pursuits of Happiness*, 176. New Hampshire's 1783 state constitution allowed for gradual emancipation. For estimates of Loyalists' slaves, see Whitfield, "Black Loyalists," 1980–1997; and idem, "Slavery in English Nova Scotia."
42. July 16, 1801, *Journals and Proceedings of the House of Assembly of Nova Scotia* (Halifax, 1801), 72. On the legal history of slavery in late eighteenth-century Nova Scotia, see Cahill, "Habeas Corpus," 179–209; Donovan, "Slaves in Île Royale," 25–42; Troxler, "Re-enslavement of Black Loyalists," 70–85; and

Gould, *Among the Powers of the Earth*, Chap. 5. For an overview of this period in pre-Confederation Canadian history, see Winks, *The Blacks in Canada*, 24–113.

43. Whitfield, "Slavery in English Nova Scotia," 23. On gradual emancipation in the North more generally, see Whitfield, *Blacks on the Border*; and Minardi, *Making Slavery History*.
44. July 30–31, 1801, Titus Smith Papers RG1, Vol. 380A, NSA. For a detailed account of grants and the amount and geography of acreage distributed to both white and black Loyalists, see Gilroy, *Loyalists and Land Settlement*.
45. "Journal of a tour . . . W. Booth," MG1 Vol. 144, NSA; Robertson, "Curatorial Report 91";and Niven, "Curatorial Report 93"; Hodges, ed. *The Black Loyalist* Directory; Frey, *Water from the Rock*.
46. *Clarkson's Mission to America, 1791–1792*, PANS Publication No. 11 (Halifax: Public Archives of Nova Scotia, 1971), 89–90.
47. JW to Portland, Oct. 29, 1796, RG 1 Vol. 51/6, 1793–1796, WP-NSA.
48. JW to Portland, Oct. 29, 1796, RG 1 Vol. 51/6, 1793–1796, WP-NSA; Dallas, *History of the Maroons*, Vol. 2, 201. Estimated numbers of emigrants are derived from Sierra Leone Company, *An Account of the Colony of Sierra Leone*, 1–3 (1,196); Jasanoff, *Liberty's Exiles*, 292 (cold place); and Byrd, *Captives and Voyagers*, 185 (more than a thousand). Among the reasons blacks gave for leaving Nova Scotia, Byrd does not include the climate, 186–190.
49. Cuthbertson, *The Loyalist Governor*, 6–8, 79; Dec. 24, 1798, JW to Edward Winslow in Raymond, ed., *Winslow Papers*, Vol. 2, 434–435 (natural history); Barry Cahill, "Mediating a Scottish Enlightenment Ideal: The Presbyterian Dissenter Attack on Slavery in Late Eighteenth-Century Nova Scotia," in Harper and Vance, eds., *Myth, Migration, and the Making of Memory*, 189–201.
50. Cuthbertson, *The Loyalist Governor*, 78.
51. Feb. 24, 1784, JW to Paul Wentworth, RG 1, Vol. 49/4, WP-NSA.
52. Bumsted, *Land, Settlement, and Politics*, 33, 44; Sylvia Hamilton, "Naming Names, Naming Ourselves: A Survey of Early Black Women in Nova Scotia," in Bristow, ed., *We're Rooted Here*, 30. In 1763, the Earl of Egmont proposed making Prince Edward Island into a northern plantation colony by importing slaves from Dominica. O'Grady, *Exiles and Islanders*, 7–8.
53. Kopytoff, "Early Jamaican Political Development"; Geggus, "The Enigma of Jamaica"; Richard Sheridan, "The Maroons of Jamaica, 1730-1830: Livelihood, Demography, and Health," in Heuman, ed., *Out of the House of Bondage*, 152–172 (quote on 154).
54. Long, *The History of Jamaica*, Vol. 3, 59.
55. June 28, 1797, *Journals and Proceedings of the House of Assembly of Nova Scotia* (Halifax, 1797), 251.
56. Hudson, "From 'Nation' to 'Race,'" 247–264; Drayton, *Nature's Government*, 92; Roberts, *Slavery and the Enlightenment*.

57. *Plan for the Abolition of Slavery in the West Indies* (London, 1772); Brown, *Moral Capital*; Sierra Leone Company, *An Account of the Colony of Sierra Leone*, 1–3.
58. Brown, *Moral Capital*, 2–5.
59. Equiano, *The Life of Olaudah Equiano*, Vol. 2, 252.
60. Brown, *Moral Capital*, 226–228; Stephen J. Braidwood, *Black Poor and White Philanthropists*, 65–67, 74–77.
61. *Plan for the Abolition of Slavery in the West Indies* (London, 1772), 5, 15–25; Brown, *Moral Capital*, 213–219; Drayton, *Nature's Government*.
62. Douglass, *A Summary Historical and Political*, 134.
63. JW to Grey Elliott, Esq., Apr. 10, 1784, Box 1, Folder 4B, Wentworth Family Papers, New Hampshire Historical Society; Gwyn, *Excessive Expectations*, 26–28.
64. Sep. 20, 1796 and Aug. 13, 1796, JW to Portland, RG 1 Vol. 51/337; Apr. 21, 1797, JW to Duke of Portland, RG 1 Vol. 52/7, WP-NSA.
65. Dallas, *History of the Maroons*, Vol. 2, 490.
66. *Eighteenth Parliament of Great Britain: Second Session* (November 2, 1797–June 29,1798), May 1, 1798, 90.
67. On the Scots' Darien adventure in Panama, see McNeill, *Mosquito Empires* and Plank, *An Unsettled Conquest*, 42, 48. For Kourou, see Rothschild, "A Horrible Tragedy," 67–108; Hodson, "'A Bondage So Harsh,'" 95–131; and idem., *Acadian Diaspora*, Chap. 3. For the colony of New Smyrna, see Bailyn, *Voyagers to the West*, 3–66. For white settlers' anxieties about extreme climate in the antipodes, see Anderson, *The Cultivation of Whiteness*.
68. Dallas, *History of the Maroons*, Vol. 2, 199–200.

CHAPTER 5

1. "List of Plans referred to in the General Description of the Province of Nova Scotia, 1783," RG 1 Vol. 284, No. 23, JDF-NSA; Jonsson, *Enlightenment's Frontier*, Chaps. 3 and 5; Golinski, *British Weather and the Climate of Enlightenment*, Chap. 6; Richard Grove, "The Evolution of the Colonial Discourse on Deforestation and Climate Change, 1500–1940," in *Ecology, Climate, and Empire*, 5–36; idem, *Green Imperialism*, Chaps. 4 and 5; Fleming, *Historical Perspectives*, 11–32.
2. Crèvecoeur, *Letters to an American Farmer*, 20–21; Ayscough, *Remarks on the Letters from an American Farmer*, 3–7; Frances Wentworth to Lady Fitzwilliam, Oct. 5, 1784, Governor John Wentworth Fonds, MG1-NSA; on Locke, see Armitage, *Foundations*, 79–131.
3. Slack, *The Invention of Improvement*, 2; Marshall and Williams, *The Great Map of Mankind*.
4. *Letters and Papers on Agriculture: Extracted from the Correspondence of a Society Instituted at Halifax for Promoting Agriculture in the Province of Nova Scotia*, Vol. 1 (Halifax: John Howe, 1789), 5 (ancestors); Jonsson, *Enlightenment's Frontier*; Paul Warde, "The Idea of Improvement, c.1520-1700," in Hoyle, ed., *Custom, Improvement,*

and the Landscape, 127–148; Drayton, *Nature's Government*; Simon Schaffer, "The Earth's Fertility as a Social Fact in Early Modern England," in Teich, Porter, and Gustafsson, eds., *Nature and Society in Historical Context*, 124–147; Chaplin, *An Anxious Pursuit*; Ritvo, *The Animal Estate*; Barrell, *The Dark Side of the Landscape*.

5. Slack, *The Invention of Improvement*; Dickson, *Old World Colony*, 170–248; Anderson, *Creatures of Empire*; Armitage, *Ideological Origins*, 49–50, 97; Cronon, *Changes in the Land*, 19–22; Canny, "The Ideology of English Colonization," 575–579.
6. Winthrop, *Reasons to Be Considered for Justifying the Undertakers of the Intended Plantation in New England*; Columella, [Reprinted letter "To the Secretary of the Agricultural Society at Halifax," March 5, 1790], *Nova Scotia Magazine*, May 1790; January 6, 1795, *The Royal Gazette and the Nova-Scotia Advertiser*; Isaac Deschamps, "Observations on the Progress of Agriculture in Nova Scotia and New Brunswick with Notices of Acadian Manners and Customs . . ." (1782), RG 1, Vol. 284, NSA.
7. McLaren, ed., *Despotic Dominion*; Belich, *Replenishing the Earth*; Yirush, *Settlers, Liberty, and Empire*.
8. "An Explanatory Supplement to Capt. Vetch's Proposal for an Attack upon Quebec and Montreal," [17 Nov.] 1708, CO 323/6, No. 71 and CO 324/9; P. Medows, J. Bruce, and J. Merrill to Board of Trade, June 22, 1717, MG1, Vol. 1520, Folder M, [copy of PRO CO 217/2, WA 53], NSA.
9. Paul Mascarene, "Description of Nova Scotia," 1720/21, MG1, Vol. 1520, Folder P [copy of PRO CO 217/3]; Council for Nova Scotia, Report to the King, September 27, 1720, MG1, Vol. 1520, Folder O, [copy of PRO CO 217/4, WA 53], NSA; Hodson, *Acadian Diaspora*, 34–35.
10. Cape Breton Governor's Accounts, 1784–1801, J.F.W. DesBarres Fonds, Series 3, 1774-1807 (MG 23, F1-3), Library and Archives,Canada.
11. *Rules and Regulations of the Massachusetts Society for Promoting Agriculture* (Boston: Fleet, 1796), 4–5; Thornton, *Cultivating Gentlemen*.
12. Young, *Annals of Agriculture*, 9–84; "A Month's Tour to Northamptonshire, Leicestershire, &c. July 18, 1791," in Young, *Tours in England and Wales: Selected from the Annals of Agriculture* (London: London School of Economics, 1932), 203–204.
13. Hugh Findlay to Oliver Smith, Dec. 4, 1793, Box 1, Folder 11 and William Prince to Oliver Smith, Apr. 14, 1794, Box 2, Folder 11, MSPA, MHS; Liancourt, *Travels through the United States of North America*, 550; Miller, *A Brief Retrospect of the Eighteenth Century*, Vol. 1, 388–389.
14. Eliot, *Essays Upon Field Husbandry in New England*, 34; Woodward, *Ploughs and Politicks*, 70; John Leverett to Aaron Dexter, Apr. 20, 1797, Box 1, and Benjamin Percival to John Lowell, March 2, 1816, Box 2, Folder 11, MSPA-MHS; Robbins, *Address to the Society for the Promotion of Agriculture*; WDP, "Report for the President of the Board of Visitors on trip to Europe," Box 2, Folder: Papers 1808–1810, WDP Papers, Harvard University Archives.

15. Peter Melançon to DesBarres, Sep.14, 1773, J.F.W. DesBarres Fonds, Series 5, Additional correspondence and papers, 1762–1894 (MG 23, F1-5); Tatamagouche and Minudie Estates, DesBarres Papers, Series 2, MG 23, F1, National Archives of Canada. On improvement in nineteenth-century Nova Scotia, see Wynn, "Exciting a Spirit of Emulation," 5–51; Samson, *The Spirit of Industry*.
16. Tatamagouche and Minudie Estates, DesBarres Papers, NAC; Robinson and Rispin, *Journey through Nova-Scotia*, 24; Isaac Deschamps, "Observations on the Progress of Agriculture in Nova Scotia and New Brunswick with Notices of the Acadians who were settled before us in Manners and Customs . . ." 1782, RG 1, Vol. 284, NSA; John Brittain, "Description of the Harbours on the coast of NS from Halifax to Mahone Bay," and William Shaw, Halifax to Colonel Edward Winslow, Jan. 1 1785, C.B. Ferguson Papers, MG 1 Box 1898, Folder 9, Typescript, NSA.
17. Tatamagouche and Minudie Estates, DesBarres Papers, NAC; Young quoted in G. E. Mingay, ed., *The Agricultural Revolution*, 11–12; Board to Mr. John Clark, March 1, 1796, RASE-B.XIII.64, MERL.
18. Timothy Ruggles to Edward Winslow, July 17, 1783, MG 100, Vol. 216, No. 15-17, NSA; William Shaw, Halifax to Colonel Edward Winslow, Feb. 1, 1785, C.B. Ferguson Papers, MG 1 Box 1898, Folder 9, NSA.
19. Liancourt, *Travels through the United States of North America*, Vol. 1, 556–558, Vol. 2, 120; Jeyes, *The Russells of Birmingham*, 174–175, 188–189, 237–241; Martha Russell, "Journal of a Tour to America, 1794-1795," in Vickery, ed. *Women's Language and Experience*.
20. Jonsson, *Enlightenment's Frontier*; Sutherland, *Taming the Wild Field*, 43–53; Koerner, *Linnaeus*, 116.
21. Dwight, *Travels in New England and New York*, Vol. 1, 83; Benjamin Vaughan to Judge Lowell, March 19, 1800, Carton 22: SH 116Y1, Folder: Misc. Agricultural Papers, Vaughan Family Papers, MHS; March–October 1811, Farm Journals, 1808–1848, Peter Chardon Brooks Papers, Ms N-2049, MHS; "On the Utility of Frost Conductors," *Nova Scotia Magazine*, June 1790; WDP to Theodore Lyman, London August 16, 1805, Box 2, Folder—Papers 1805-07, HUG 1677, WDPP; William Aiton to Joseph Banks [October 4, 1803], BL Addl. Mss. 33982.264–265.
22. Deane, *The New-England Farmer*, 85, 88; Eliot, *Essays Upon Field Husbandry in New England*, 31; *Transactions of the Society for Promoting Agriculture in the State of Connecticut* (New Haven: William W. Morse, 1802), 3; Sarah Vaughan to Charles Vaughan, Sep. 1797, Carton 1, VFP, MHS; William Manning to Sarah Vaughan, Feb. 1798, Carton 1, VFP, MHS.
23. Aug. 16, 1766, Samuel Holland to John Pownal, CO 323/24.46-47, PRO; Pratt, *Imperial Eyes*, esp. 15-37; Sherman, "Stirrings and Searchings," 29–32; Hubbard Jr., *American Boundaries*, 20–23.
24. Spary, *Utopia's Garden*, 13; [Edward Long], *A Free and Candid Review of a Tract, entitled 'Observations on the Commerce of the American States* (London,

1784), 10; "Lecture: Introductory on Natural History, October 12, 1810," Folder c. 16.4, BW-HMS; *Letters and Papers on Agriculture* (Halifax: John Howe, 1789), 5–6; [Draft of inaugural address to Harvard Corporation] [1805], Folder—Papers 1805-07, Box 2 HUG 1677, WDPP; "Lecture: Introductory on Natural History," BW-HMS; Titus Smith, "Notes on Botany," Folder 5, item 24, MG1 Vol. 1664B, NSA.

25. White, *The Natural History*; June 4, 1801, William Sabatier, 'Circular on behalf of the compilers of a history,' Vol. 15, Lt. Governor Prince Edward Island, 1804–1812, J.F.W. DesBarres Fonds, NAC; Belknap, *History of New-Hampshire,* Vol. 3, 3.
26. July 27–28, 1731, Robert Hale's Journal of an Expedition to Nova Scotia, 1731, (typescript), MG1, Box 1898, Folder 10.1, NSA; Nov. 21, 1715, Thomas Caulfield to Board of Trade, MG1 Vol. 1520, Folder K, NSA; Titus Smith, *Survey of the Eastern and Northern Parts of the Province.*
27. Bailyn, *Voyagers to the West*, 10–11; Belknap, *The History of New-Hampshire*, Vol. 3: 42, 47, 54, 97; Lawson, *Passaconaway's Realm.* Subsequent surveys were undertaken in 1804 and 1816.
28. Belknap, "Description of the White Mountains in New-Hampshire, 42–49, quotes on 43, 46; Cutler and Cutler, eds., *Life, Journals and Correspondence*, Vol. 2, 98–99, 103; Belknap, *The History of New-Hampshire*, Vol. 3, 48–51; Belknap, *Journal of a Tour to the White Mountains*, 16; Alexander, "The Presidential Range of New Hampshire," 104; Bermingham, *Landscape and Ideology*; and MacLaren, "The Limits of the Picturesque," 97–111.
29. "Change of Climate in North America and Europe," n.d. Box 1, Folder 10, Samuel Williams Papers, Harvard University Archives; "Natural History" Folder 65, Samuel Williams Family Papers, Ms AM 2624, Houghton Library; Williams, *Natural and Civil History of Vermont*, 42–65; Golinski, "American Climate and the Civilization of Nature," in Dew and Delbourgo, *Science and Empire in the Atlantic World*, 153–174; Judd, *Untilled Garden*, 228–235; Fleming, *Historical Perspectives*, 11–32.
30. Glacken, *Traces on the Rhodian Shore*; Adolph B. Benson, ed. and trans., *Peter Kalm's Travels in North America* (New York: Wilson-Erickson, 1964); Jonsson, "Climate Change and the Retreat of the Atlantic," 99–126; Barnett, "The Theology of Climate Change," 217–237.
31. John Smith, *A Description of New England*, 14–15 (men of industrie); *Journal of John Winthrop*, 35; Vincent, *A True Relation of the Late Battle Fought between the English and the Pequot Savages*; Thwaites, ed., *The Jesuit Relations and Allied Documents*, Vol. 2, 201, Vol. 4, 111; Talon, "Mémoire sur l'état présent du Canada," 63; Kupperman, "Climate and Mastery of the Wilderness," 3–38. Kupperman argues that a decline in this hopeful view in late seventeenth-century sources reflects the coincidence of particularly harsh winters during the Little Ice Age and Puritan ministers' anxiety about the declining numbers and rigorous piety of their congregations. From the perspective of a broader chronological scope, however, this

argument loses its persuasiveness because historical climatology largely fails to match early American naturalists' ongoing beliefs in climate warming.

32. Vogel, "The Letter from Dublin," 111–128; Hume, "Discourse X: Of the Populousness of Ancient Nations," in *Political Discourses*, 246; Williamson, "An Attempt to Account for the Change of Climate," 277–280; "Notes on the State of Virginia" (Manuscript, 1781–1785), Thomas Jefferson Papers, MHS; Ramsay, *History of South Carolina*, Vol. 2, 64-69, 304; Anon., *American Husbandry*, 2 (bad), 46 (amelioration).
33. "Thermometrical Register (1763–1795)," Ezra Stiles Papers, Beinecke Library, Yale University; Ezra Stiles to Samuel Williams, June 5, 1782, Box 1, Folder 32, Samuel Williams Papers, Special Collections, University of Vermont Library; Cutler, "Meteorological Observations," 336–371; "Vermont No. 1," Folder 67, Samuel Williams Family Papers, Ms AM 2624, Houghton Library (DesBarres); Anon. "Meteorological Observations Made at Montreal, Canada, North America," *Memoirs of the American Academy of Arts and Sciences* 2: 1 (1793), 113–118; Williams, *Natural and Civil History of Vermont*, 57–60; BV to Nathaniel Dummer, Captain Randall, and Captain Grant, Select-men of the town of Hallowell, Oct. 4, 1798, VFP-MHS. Slonosky, "Meteorological Observations," 2232–2247; Morgan, *The Gentle Puritan*, 134–136.
34. Williams, *Natural and Civil History of Vermont*, 57 (new country); House of Commons, "Report from the Committee on the Petition of the Court of Directors of the Sierra Leone Company," (London, 1802), 744; R. Philipps to Board of Trade, "An Account of the situation in Nova Scotia, Nov. 26 1730, MG1, Vol. 1520, Folder S, NSA; *A Genuine Account of Nova Scotia To which is added, His Majesty's Proposals, as an encouragement to those who are willing to settle there* (London, 1750), 4. This pamphlet was reprinted from the *London Magazine* (March 1749) and also circulated in Dublin.
35. Mitchell, *Present State of Great Britain and North America*, Vol. 1, 167; [Edward Long], *A Free and Candid Review of a Tract, entitled 'Observations on the Commerce of the American States* (London, 1784), 15–16, 21, 44; [John Holroyd Sheffield], *Observations on the Commerce of the American States with Europe and the West Indies* (London, 1783). For Priestley and aerial reform, see Golinski, *British Weather*, 159–166.
36. John Mervin Nooth to Joseph Banks, *Recueil de lettres autographes signées du docteur J. Mervin Nooth*, 1789–1799, 1902, *Ville de Montréal—Section des archives* (fictive Quebec volcano); Zilberstein, "Inured to Empire," 125–156. On earthquakes and climate change, see Jonsson, *Enlightenment's Frontier*, Chap. 3; and Wood, *Tambora*
37. Rudwick, *Bursting the Limits of Time*, 75–80; Buffon, *Des époques*, 1 ("*dans l'histoire naturelle, il faut fouiller les archives du monde*"), 67 and 242 (age of earth), 225 ("*Lorsque la puissance de l'homme a secondé celle de la nature*"), 237 ("*enfin la face entière de la terre porte aujourd'hui l'empreinte de la puissance de l'homme,*

laquelle, quoique subordonnée à celle de la nature, souvent a fait plus qu'elle, ou du moins l'a si merveilleusement secondée, que c'est à l'aide de nos mains qu'elle s'est développée dans toute son étendue, et qu'elle est arrivée par degrés au point de perfection et de magnificence où nous la voyons aujourd'hui").

38. Leslie, "On Heat and Climate," 19–21; Leslie, *An Experimental Inquiry*, 181–182, 536–537. "On Heat and Climate" was read aloud at a meeting of the Royal Society in February or March 1793 but Banks declined to publish it in the *Philosophical Transactions*. In 1819, Leslie published the paper in its original version in the Scottish journal *Annals of Philosophy*.
39. Benjamin Franklin to Ezra Stiles, May 29, 1763, in Labaree, ed.,*The Papers of Benjamin Franklin*, Vol. 10, 264–267; Webster, "A Dissertation on the Supposed Change in the Temperature," Vol. 1, 1–68; Anon., *An Essay on the Climate of the United States* (Philadelphia, 1809); Dwight, *Travels in New England and New York*, Vol. 1, 63–65.
40. "Change of Climate in North America and Europe," n.d. Box 1, Folder 10, Samuel Williams Papers, Harvard University Archives; Dwight, *Travels in New England and New York*, Vol. 2, 14.
41. Dwight, *Travels in New England and New York*, Vol. 1, 9–12, 19–21, 86, 104–105, Vol. 2, 142, 213, 340; Kamensky, " 'In These Contrasted Climes, How Chang'd the Scene,' " 80–108.
42. Webster, "A Dissertation," quotes on 1, 68; Williams, *Natural and Civil History of Vermont*, 58; "Change of Climate in North America and Europe," n.d. Box 1, Folder 10, Samuel Williams Papers, Harvard University Archives.

Bibliography

MANUSCRIPTS

Britain

British Library, London

Additional MSS 8096.527, 8096.550, 8097.358, 8098.305-306, 33982.264-265

British Museum (Natural History)

Sir Joseph Banks Correspondence, Dawson Turner Copies

Edinburgh University Library, Special Collections

Papers of Professor Andrew Brown (1763–1834)

Georgian Archive, Royal Archives, Windsor

Augusta, Princess of Wale's Accounts, RA GEO, Box 5: Nos. 55422-55661

The Museum of English Rural Life, The University of Reading

Royal Agricultural Society of England, SR RASE BI-IV (1793–1822)

National Archives, Public Record Office, Kew

CO 5, CO 194/5, CO 217-218, CO 323/6, CO 388

National Library of Scotland

Royal Highland & Agricultural Society of Scotland Papers, Adv. MS.73.2.13, f.27

Royal Botanic Gardens, Kew, Library, Art, and Archives

Banks Collection of Original Manuscripts, BC 2.282

Kew Record book, Vol. 1, 1793–1809, Vol. 2, 1804–1826

Royal Institution of Great Britain, Archives

Benjamin Thompson, Count von Rumford Collection, GB 0116

CANADA

National Archives of Canada

Joseph Frederick W. DesBarres Papers, 1762–1894, MG23-F1

Nova Scotia Museum

Prescott Family Papers

Nova Scotia Archives

MG 1, Vols. 3–5: Akins Family Documents, 1762–1934
MG 1, Vol. 11: Almon Family Papers
MG 1, Vol. 144: Lieutenant William Booth Papers
MG 1, Vol. 249; Vol. 3335: Cunningham Papers
MG 1, Vol. 427: Harrison Family Papers
MG 1, Vol. 438: John Huston Diaries 1787–1788, 1793–1801
MG 1, Vol. 939: Governor John Wentworth Fonds
MG 1, Vol. 1520
MG 1, Vol. 1664: Titus Smith Papers
MG 1, Vol. 1773: Miscellaneous Correspondence
MG 1, Box 1898: C. B. Ferguson Papers
MG 1, Vol. 3335: Duke of Kent, Windsor, Correspondence
MG 6, Vol. 2: Minutes for Kinds County Agricultural Society, c. 1789
MG 12, No. 1-59: Royal Engineers Fonds
MG 100, Vol. 84: Susan Dalrymple
MG 100, Vol. 100: Acadians: Concerning Their Treatment in Massachusetts, 1755
MG 100, Vol. 130: Agricultural Improvement
MG 100, Vol. 134: Description of Acadian Diking Methods
MG 100, Vol. 143: "Expulsion of the Acadians"
MG 100, Vol. 169
MG 100, Vol. 216: Timothy Ruggles
RG 1, Vol. 49-57: Wentworth Papers
RG 1, Vol. 284: Transcripts of Documents at British Library Relating to Nova Scotia Affairs, 1750–1790
RG 1, Vol. 364: John Allan Journal
RG 1, Vol. 380: Titus Smith Tours
RG 5, Series P, Vol. 51: Petitions to the House of Assembly, 1816–1828
RG 20, Series C, Vol. 34: Crown Lands, Hants County, 1763–1839

Ville de Montréal—Section des archives

Recueil de lettres autographes signées du docteur J. Mervin Nooth, 1789–1799, 1902

UNITED STATES

The Gray Herbarium, Botany Libraries Archives, Harvard University

Cutler / Dandridge / Thorndike Papers
Manasseh Cutler Papers, 1742–1823

Countway Library, Harvard Medical School
Benjamin Waterhouse Papers, H MS c16.4

Harvard University Archives
Papers of Samuel Williams, 1752–1794, HUM 8
William Dandridge Peck Papers, HUG 1677

Houghton Library, Harvard University
Samuel Williams Family Papers, Ms AM 2624

Huntington Library and Archives, San Marino, CA
Appleton-Foster Collection, HM 27635 277OS
Spence-Lowell Collection, SL1-317

John Hay Library, Brown University
Drowne Family Papers, Ms. Drowne

Massachusetts Historical Society
Brooks Family Papers, 1751–1861 Ms. N-1948, Ms. N-163
Samuel Chandler Diary, 1746–1772, microform P-7
Charles Howland Papers, 1784–1836, Ms N-2203
Massachusetts Society for Promoting Agriculture Papers, Ms. N-517
Ebenezer Parkman Papers, 1749, 1755, 1771-73, microfilm P-363, reel 7.3-7.5
— Commonplace Book, 1771–1773, 1781–1782, microfilm P-220
Joseph Shaw Papers, Ms N-918
Thaxter Family Papers, Ms-N-1655
Vaughan Family Papers, Correspondence, 1773–1812, Ms. SH 116, Cartons 1, 2, 22
Thomas Waldron, Ms. S-758
Phineas Whitney Papers, Ms. S-246
Aaron Wight Diary, microfilm P363-reel 11

New Hampshire Historical Society
Jeremy Belknap Papers
Wentworth Family Papers

Rare Book and Manuscript Library, Columbia University
Samuel Vetch papers, 1708–1712, MS#1292

University of Vermont Library, Special Collections
Samuel Williams Papers

Yale University Library, Manuscripts and Archives
Morse Family Papers, MS 358

Beinecke Library, Yale University
Ezra Stiles Papers, MS Vault Stiles, MTJ 1-6

PERIODICALS

Acadian Recorder
Columbian Centinel

Halifax Gazette
London Magazine
New-England Weekly Journal
Nova Scotia Gazette
Nova Scotia Gazette and Weekly Chronicle
Nova Scotia Magazine and Comprehensive Review of Literature, Politics, and News
The Nova Scotia Chronicle
The Nova Scotia Chronicle and Weekly Advertiser
The Royal Gazette and the Nova-Scotia Advertiser
Scots Magazine

PRINTED PRIMARY SOURCES

Adams, John. *Statesman and Friend: Correspondence of John Adams with Benjamin Waterhouse, 1784–1822*. Worthington C. Ford, ed. Boston: Little, Brown, 1927.

A Genuine Account of Nova Scotia To which is added, His Majesty's Proposals, as an encouragement to those who are willing to settle there. London, 1750.

Alexander, William. *An Encouragement to Colonies*. London, 1624.

American Husbandry: Containing an Account of the Soil, Climate, Production, and Agriculture. London: J. Bew, 1775.

Ames, Nathaniel. *An Astronomical Diary, or An Almanack ... Calculated for the Meridian of Boston in New-England*. Boston, 1758–1775.

"An Account of a Sort of Sugar Made of the Juice of the Maple, in Canada." *Philosophical Transactions* 15 (1685), 988.

Anderson, Wallace, ed. *The Works of Jonathan Edwards: Vol. 6: Scientific and Philosophical Writings*. New Haven, CT: Yale University Press, 1980.

A New Collection of Voyages, Discoveries and Travels, Containing Whatever is Worthy of Notice. 7 vols. London, 1767.

An Essay on the Climate of the United States. Philadelphia, 1809.

Arbuthnot, John. *Essay Concerning the Effects of the Air on Human Bodies*. London, 1733.

A True Account of the Colonies of Nova Scotia and Georgia. Newcastle: Printed in the present year [1780?].

Ayscough, Samuel. *Remarks on the Letters from an American Farmer*. London, 1783.

Belknap, Jeremy. *The History of New-Hampshire*. 3 vols. Boston: 1784.

—— "Description of the White Mountains in New-Hampshire, Read October 15, 1784." *Transactions of the American Philosophical Society* 2 (1786), 42–49.

—— "On the Preserving of Parsnips by Drying." *Transactions of the American Philosophical Society* 2 (1786), 199–200.

—— *A Discourse, Intended to Commemorate the Discovery of America by Christopher Columbus; delivered at the request of the Historical Society*. Boston, 1792.

—— "Journal of a Tour to the White Mountains in July, 1784." Boston: Massachusetts Historical Society, 1876.

—— *Belknap Papers*. Series 6. Vol. 4. Boston: Massachusetts Historical Society, 1891, 521–615.

Best, George. *The Three Voyages of Martin Frobisher*. London: Hakluyt Society, 1867.

Betham-Edwards, M., ed. *The Autobiography of Arthur Young*. London, 1898.

Biard, Pierre. *Relation de la Nouvelle France, de ses terres, naturel du pais, & de ses habitants*. Lyon, 1616.

Bigelow, Jacob. *Florula bostoniensis: A Collection of Plants of Boston and its Environs*. Boston: Cummings and Hilliard, 1814.

Biggar, H. P. *The Voyages of Jacques Cartier*. Toronto: Toronto University Press, 1993.

Black, William. *A Comparative View of the Mortality of the Human Species*. London, 1788.

Bodin, Jean. *Method for the Easy Comprehension of History*. Beatrice Reynolds, trans. New York: Columbia University Press, 1945.

Borland, Francis. *Memoirs of Darien*. Glasgow, 1715.

Bourne, Edward G., ed. *The Voyages and Explorations of Samuel de Champlain (1604–1616)*. Annie N. Bourne, trans. Toronto: Courier Press, 1911.

Boyle, Robert. *General Heads for the Natural History of a Country Great or Small Drawn Out for the Use of Travellers and Navigators*. London, 1691.

—— *General History of the Air*. London, 1692.

Bradford, William. *A Relation or Iournall of the Beginning and Proceedings of the English Plantation Setled at Plimoth in New England*. London, 1622.

—— *History of Plymouth Plantation, 1606–1646*. William T. Davis, ed. New York: Scribner, 1908.

Brereton, John. *A Briefe and True Relation of the Discouerie of the North Part of Virginia*. London, 1602.

Bridenbaugh, Carl, ed. *Gentleman's Progress: The Itinerarium of Dr. Alexander Hamilton, 1744*. Chapel Hill: University of North Carolina Press, 1948.

Brown, Sanford C., ed. *Collected Works of Count Rumford, Vol. 1: The Nature of Heat*. Cambridge,MA: Harvard University Press, 1968.

Browne, Peter. *The Civil and Natural History of Jamaica in Three Parts*. London, 1756.

Brunhouse, Robert L. "David Ramsay, 1749–1815: Selections From His Writings." *Transactions of the American Philosophical Society* 55(4) (1965), 1–250.

Buffon, George-Louis Leclerc. *Natural History of the Horse*. London, 1762.

—— *Histoire Naturelle: générale et particulière. Des époques de la nature*. Vol. 5. Paris, 1778.

—— *Natural History, General and Particular*. W. Smellie, trans. London: Strahan and Cadell, 1781 (1761).

Burke, William. *An Account of the European Settlements in America*. 2 vols. London, 1757.

Burrage, Henry S., ed. *Early English and French Voyages, Chiefly from Hakluyt 1534–1608*. New York: Barnes & Noble, 1952.

Carman, Harry J., ed., *American Husbandry*. New York: Columbia University Press, 1939.

Censuses of Canada, 1665 to 1871, Statistics of Canada, Vol. 4, accessed online: http// www.statcan.gc.ca/pub/98-187-x/4064810-eng.htm.

Chambers, Neil, ed. *The Letters of Sir Joseph Banks: A Selection, 1768–1820*. London: Imperial College Press, 2000.

Clarkson's Mission to America, 1791–1792. PANS Publication No. 11. Halifax: Public Archives of Nova Scotia, 1971.

Clifton, Francis, ed. *Hippocrates Upon Air, Water, and Situation*. London, 1734.

The Correspondence of the Right Honourable Sir John Sinclair. 2 vols. London: Henry Colburn and Richard Bentley, 1831.

Crèvecoeur, J. Hector St. John de. *Letters to an American Farmer*. London, 1782.

Cutler, Manasseh. "An Account of Some of the Vegetable Productions, Naturally Growing in this Part of America, Botanically Arranged." *Memoirs of the American Academy of Arts and Sciences* 1 (1783), 395–493.

—— *An Explanation of the Map . . . of the Federal Lands Comprehended between Pennsylvania west line, the rivers Ohio and Sioto, and Lake Erie*. Salem, MA: Dabney and Cushing, 1787.

—— "Meteorological Observations at Ipswich in 1781, 1782, and 1783, Lat. 42° 38' 30" N Long. 70° 45' W." *Memoirs of the American Academy of Arts and Sciences* 1 (1783), 336–371.

Cutler, William and Julia Cutler, eds. *Life, Journals, and Correspondence of Reverend Manasseh Cutler, LLD, By His Grandchildren*. 2 vols. Cincinnati: Robert Clarke & Co., 1888.

D'Abbeville, Claude. *Histoire des Pères Capucins en l'Isle de Maragnan et terres circonvoisines*. Paris, 1614.

Dallas, Robert C. *The History of the Maroons, From Their Origin to the Establishment of Their Chief Tribe at Sierra Leone*. 2 vols. London: Longman, 1803.

Davenant, Charles. *An Essay upon Ways and Means of Supplying the War*. London, 1701.

—— *Some Objections against Settling the Trade to Africa*. London, 1709.

—— *Reflections upon the Constitution and Management of the Trade to Africa*. London, 1709.

—— *An Essay upon Universal Monarchy, Written in the year 1701*. London, 1734.

Davis, John. *The Seaman's Secrets, Divided into Two Parts*. London, 1599.

Deane, Samuel. *The New-England Farmer, or, Georgical Dictionary*. Worcester, MA: Isaiah Thomas, 1790.

Denys, Nicholas. *Description géographique et historique des costes de l'Amérique septentrionale: avec l'histoire naturelle du païs*. Paris, 1672.

DesBarres, Joseph Frederick Wallet. *The Atlantic Neptune*. London, 1781. http://www.nmm.ac.uk/collections/explore/object.cfm?ID=HNS%20139.

Dickson, Adam. *A Treatise of Agriculture: A New Edition*. 2 vols. Edinburgh, 1785 [1762].

Douglass, William. *A Summary Historical and Political, Of the First Planting, Progressive Improvements, and Present State of the British Settlements in North America*. 2 vols. Boston: Rogers and Fowle, 1747–1752.

Dudley, Paul. "Observations on Some of the Plants in New England." *Philosophical Transactions of the Royal Society* 33 (1724–1725), 194–200.

Dunn, Richard S. and Laetitia Yeandle, eds. *The Journal of John Winthrop, 1630–1649*. Cambridge, MA: Harvard University Press, 1996.

Dwight, Timothy. *Travels in New England and New York*. 4 vols. New Haven, CT, 1821.

Eliot, Jared. *Essays Upon Field Husbandry in New England, As It Is Or May Be Ordered*. Boston: Edes & Gill, 1760.

—— *Essays Upon Field-Husbandry in New England and Other Places, 1748-62*. Harry Carman et al., eds. New York: Columbia University Press, 1934.

Equiano, Olaudah. *The Interesting Narrative of the Life of Olaudah Equiano, or Gustavus Vassa, the African. Written by Himself*. 2 vols. London, 1789.

Extracts and Copies of Letters from Sir John Wentworth, Lieutenant Governor of Nova Scotia, to his Grace the Duke of Portland, respecting the Settlement of the Maroons in that Province. London, 1797.

Franklin, James. *The Philosophical and Political History of the Thirteen United States of America*. London, 1784.

Gibbon, Edward. *The History of the Decline and Fall of the Roman Empire*. London, 1776.

Gillies, John, ed. *Aristotle's Ethics and Politics*. 2 vols. London, 1797.

Goldsmith, Oliver. *The History of the Earth*. 8 vols. London, 1774.

Gorges, Ferdinando. *Briefe Narration of the Original Undertakings of the Advancement of Plantations into the Parts of America*. London, 1622.

Hale, Samuel. "Conjectures of the Natural Causes of the North West Winds Being Colder, and More Frequent in the Winter in New England, than in the Same Degrees of Latitude in Europe." *Memoirs of the American Academy of Arts and Sciences* 2:1 (1793), 61–63.

Hales, Stephen, *Vegetable Staticks*. London, 1727.

Haliburton, Robert Grant. *The Men of the North and Their Place in History*. Montreal, 1869.

Hawley, Stephen. "The Agency of God in Snow, Frost, & c." New Haven, CT, 1771.

Hill, John. *Eden, or, A compleat body of gardening*. London, 1773.

Holyoke, Samuel. "An Estimate on the Excels of the Heat and Cold of the American Atmosphere Beyond the European." *Memoirs of the American Academy of Arts and Sciences* 2:1 (1793), 65–88.

House of Commons. "Report from the Committee on the Petition of the Court of Directors of the Sierra Leone Company." London, 1802.

Howard, Luke. *On the Modification of Clouds*, 3rd ed. London, 1864 (1803).

Hubbard, William. *A General History of New England: From the Discovery to 1680*. Boston: Massachusetts Historical Society, 1815.

Hume, David. *Political Discourses*. Edinburgh, 1752.

Hutchinson, Thomas. *The History of the Colony of Massachuset's Bay*. Boston, 1764.

James, Sydney V., ed. *Three Visitors to Early Plymouth: Letters About the Pilgrim Settlement in New England During its First Seven Years*. Bedford, MA: Applewood Books, 1997.

Jefferys, Thomas. *The Natural and Civil History of the French Dominions in North and South America*. London, 1760.

Jeyes, S.H. *The Russells of Birmingham in the French Revolution and in America, 1791–1814*. London: George Allen, 1911.

Johnson, Samuel. "Climate." *A Dictionary of the English Language*. 2nd ed. London, 1760.

Josselyn, John. *New England's Rarities Discovered*. London, 1672.

Journals and Proceedings of the House of Assembly of Nova Scotia. Halifax, 1797.

Journals and Proceedings of the House of Assembly of Nova Scotia. Halifax, 1801.

Kalm, Pehr. *Travels into North America*. 2 vols. John Reinhold Forster, trans. Warrington, England, 1770.

Klockhoff, H. "A Chorographical Map of the Northern Department of North America." Amsterdam, 1780.

Knight, Fred, ed. *Letters on Agriculture from His Excellency George Washington, President of the United States, to Arthur Young*. Washington, D.C., 1847.

Labaree, Leonard W., ed. *The Papers of Benjamin Franklin*. Vol. 10. New Haven, CT: Yale University Press, 1959.

Laut, Agnes C. *Canada: The Empire of the North*. Boston: Ginn and Company, 1909.

Laws and Regulations of the Massachusetts Society for Promoting Agriculture: With Extracts of Foreign and Domestic Publications. Boston: Thomas Andrews, 1793.

Leslie, John. *An Experimental Inquiry into the Nature and Propagation of Heat*. London, 1804.

—— "On Heat and Climate," *Annals of Philosophy* 14 (1819), 5–27.

Letters and Papers on Agriculture: Extracted from the Correspondence of a Society Instituted at Halifax for Promoting Agriculture in the Province of Nova Scotia. Halifax: John Howe, 1789.

Liancourt, Duke de la Rochefoucauld. *Travels through the United States of North America, The Country of the Iroquois, and Upper Canada, in the Years 1795, 1796, and 1797*. H. Neuman, trans. 2nd ed. 4 vols. London, 1800.

Lind, James. *Essay on Diseases Incidental to Europeans in Hot Climates*. London, 1768.

[Little, Otis]. *The State of Trade in the Northern Colonies Considered, with an Account of their Produce and a Particular Description of Nova Scotia*. London, 1748.

Long, Edward. *The History of Jamaica: Or, General Survey of the Antient and Modern State of that Island*. 3 vols. London, 1774.

—— *Free and Candid Review of a Tract entitled "Observations on the Commerce of the American States with Europe and the West Indies."* London, 1784.

Marshall, William. *Proposals for a Rural Institute*. London, 1799.

Mather, Cotton. *Bonifacius: Essays to Do Good*. Boston, 1710.

"Meteorological Observations Made at Montreal, Canada, North America." *Memoirs of the American Academy of Arts and Sciences* 2(1) (1793), 113–118.

Miller, Samuel. *A Brief Retrospect of the Eighteenth Century*. 2 vols. New York, T. & J. Swords, 1803.

Mitchell, John. "An Essay upon the Causes of the Different Colours of People in Different Climates." *Philosophical Transactions* 43 (1744–1745), 102–150.

—— *A Map of the British and French Dominions in North America*. London, 1755.

—— *Present State of Great Britain and North America*. London, 1767.

Moll, Herman. *A System of Geography*. London, 1701.

Montesquieu, Charles de Secondat. *Spirit of the Laws*. 2 vols. London, 1750.

Morse, Jedediah. *American Geography: Or, A View of the Present Situation of the United States of America*. London, 1789.

—— *American Geography: Or, A View of the Present Situation of the United States of America*. 2nd ed. London, 1792.

—— *The History of America, in Two Books*. Philadelphia, 1795.

Morton, Thomas. *New English Canaan or New Canaan*. Amsterdam, 1637.

Moxon, Joseph. *A Brief Discourse of a Passage by the North-Pole to Japan, China, &c.* London, 1674.

—— *Mathematicks Made Easie: or a Compleat Mathematical Dictionary*. London, 1701.

Neal, Daniel. *History of New England*. 2 vols. London, 1720.

New England Women and the Families in the 18th and 19th Centuries. [microfilm] SERIES B: Manuscript Collections from the Newport Historical Society, 10: 0314.

Niven, Laird. "Curatorial Report 93. Was This the Home of Stephen Blucke? The Excavation of AkDi-23, Birchtown, Shelburne County." Halifax: Nova Scotia Museum, 2000.

Oldmixon, John. *The British Empire in America*. London, 1708.

Papers Relative to the Settling of the Maroons in His Majesty's Province of Nova Scotia. London, 1798.

The Parliamentary Register; or Debates and Proceedings of the House of Commons. London, 1798.

Payson, Phillips, *A Sermon Preached before the Honorable Council, and the Honorable House of Representatives, of the State of Massachusetts-Bay, in New England, at Boston*. May 27, 1778.

Peck, William D. *A Catalogue of American and Foreign Plants, Cultivated in the Botanic Garden, Cambridge, Massachusetts*. Cambridge, MA: Hilliard and Metcalf, 1818.

—— *The Natural History of the Slug Worm*. Boston: Young and Minns, 1799.

Pennant, Thomas, *Arctic Zoology*. 1st ed. London: Henry Hughes, 1785.

—— *Arctic Zoology*. 2nd ed. London: Robert Faulder, 1792 [1787].

[Pennecuik, Alexander.] *Caledonia Triumphans: A Panegyrick to the King*. Edinburgh, 1699.

Peter Kalm's Travels in North America. Adolph B. Benson, ed. and trans. New York: Wilson-Erickson, 1964.

Petty, William. *Political Arithmetick*. London, 1691.

Pickering, Charles. "On the Geographical Distribution of Plants." *Transactions of the American Philosophical Society* 3 (1831), 274–284.

—— *The Races of Man and Their Geographical Distribution*. Philadelphia, 1848.

—— *The Races of Man and Their Geographical Distribution*. Philadelphia, 1854.

Pickering, John. *A Vocabulary, or Collection of Words and Phrases Which Have Supposed to be Peculiar to the United States of America*. Cambridge, MA, 1816.

Plan for the Abolition of Slavery in the West Indies. London, 1772.

Present Conduct of the Chieftains and Proprietors of the Lands in the Highlands of Scotland. Edinburgh, 1773.

The Proceedings of the Governor and Assembly of Jamaica, in regard to the Maroon Negroes. London, 1796.

"Queries Respecting the Slavery and Emancipation of Negroes in Massachusetts, Proposed by the Hon. Judge Tucker of Virginia, and Answered by the Rev. Dr. Belknap." *Collections of the Massachusetts Historical Society*. Boston: Massachusetts Historical Society, 1835 [1795].

Quincy, Josiah. *The History of Harvard University*. Vol. 2. Cambridge, MA: John Owen, 1840.

Quinn, Daniel B. and Alison M., eds. *The English New England Voyages, 1602–1608*. London: Hakluyt Society, 1983.

Ramsay, David. *History of South Carolina*. 2 vols. Charleston, SC, 1809.

Raymond, W.O., ed. *Winslow Papers, 1776–1826*. 2 vols. St. John: New Brunswick Historical Society, 1901.

Robbins, Asher. *An Address to the Society for the Promotion of Agriculture and other Useful Arts in the State of Rhode-Island and Providence Plantations*. Newport, RI, 1802.

Robertson, Carmelita. "Curatorial Report 91. Black Loyalists of Nova Scotia: Tracing the History of Tracadie Loyalists, 1776–1787." Halifax: Nova Scotia Museum, 2000.

Robertson, William. *The History of America*. Philadelphia, 1799.

Robinson, John and Thomas Rispin. *Journey Through Nova-Scotia Containing a Particular Account of the Country and its Inhabitants*. York, 1774.

Rogers, Robert. *A Concise Account of North America: Containing a Description of the Several British Colonies on that Continent, including the Islands of Newfoundland, Cape Breton, &c*. London, 1765.

—— *A Concise Account of North America*. Dublin, 1769.

Ruggles, Henry S. *General Timothy Ruggles, 1711–1795*. Privately published, 1897.

Rules and Regulations of the Massachusetts Society for Promoting Agriculture. Boston: Fleet, 1796.

Seasonable Advice to the Landholders and Farmers in Scotland. Edinburgh, 1770.

[Sheffield, John Holroyd.] *Observations on the Commerce of the American States with Europe and the West Indies*. London, 1783.

Sierra Leone Company. *An Account of the Colony of Sierra Leone From its First Establishment in 1793: Being the substance of a report delivered to the proprietors*. London, 1795.

Smeathman, Henry. *Plan of a Settlement to be Made Near Sierra Leona on the Grain Coast of Africa*. London, 1786.

Smith, James. *A Selection of the Correspondence of Linnaeus and other Naturalists*. Vol. 2. New York: Arno Press, 1978.

Smith, John. *A Description of New England*. London, 1616.

Smith, Titus. *Survey of the Eastern and Northern Parts of the Province in the Years 1801 & 1802*. 3rd ed. Halifax: 1857.

Stefansson, Vilhjalmur. *The Northward Course of Empire*. New York: Harcourt, 1922.

Strickland, William. *Journal of a Tour in the United States of America, 1794–1795*. J. E. Strickland, ed. New York: New York Historical Society, 1971.

Sullivan, James. *History of the District of Maine*. Boston, 1795.

Talon, Jean. "Mémoire sur l'état présent du Canada, 1667." *Rapport de l'Archiviste de la Province de Québec*. Québec: *Imprimerie du roi*, 1930–1931.

Thwaites, Reuben Gold, ed. *The Jesuit Relations and Allied Documents: Travels and Explorations of the Jesuit Missionaries in New France, 1610-1791*. Cleveland: Burrows Bros. Co., 1896–1901.

Transactions of the Society for Promoting Agriculture in the State of Connecticut. New Haven, CT: William W. Morse, 1802.

Tyndall, John. *Lectures on Count Rumford, Originator of the Royal Institution, delivered May 3, 10, 17, 1883*. London: Royal Institution of Britain, 1883.

Vetch, Samuel. "A Poem, Humbly Dedicated to the Queen Upon Her Birth-Day." London, 1707.

Vickery, Amanda, ed. *Women's Language and Experience, 1500–1940: Women's Diaries and Related Sources* [microfilm]. Marlborough, England: Adam Matthew Publications, 1994–2004.

Vincent, Philip. *A True Relation of the Late Battell Fought Between the English and the Salvages*. London, 1637.

—— *A True Relation of the Late Battle Fought between the English and the Pequot Savages . . . With the Present State of Things There*. London, 1638.

Volney, C. F. A *View of the Soil and Climate of the United States of America*. London, 1804.

Waterhouse, Benjamin. *Heads of a Course of Lectures on Natural History, Now Delivering in the University at Cambridge*. Providence, RI: Bennett Wheeler, 1784–1791.

—— *Heads of a Course of Lectures on Natural History*. Cambridge, MA: Hilliard & Metcalf, 1810.

—— *The Botanist: Being the Botanical Part of a Course of Lectures on Natural History*. Boston: Joseph T. Buckingham, 1811.

Webster, Noah. "A Dissertation on the Supposed Change in the Temperature of Winter (Read before the Connecticut Academy of Arts and Sciences, 1799)." *Memoirs of the Connecticut Academy of Arts and Sciences* 1 (1810), 1–68.

White, Gilbert. *The Natural History and Antiquities of Selborne*. London, 1789.

Williams, Samuel. *The Natural and Civil History of Vermont*. Rutland, VT, 1794.

Williamson, Hugh. "An Attempt to Account for the Change of Climate, Which has Been Observed in the Middle Colonies of North-America." *Transactions of the American Philosophical Society* 1 (January 1, 1769–January 1, 1771), 277–280.

Winslow, Edward. *Good Newes From New England*. London, 1624.

Winthrop, John. *Reasons to be Considered for Justifying the Undertakers of the Intended Plantation in New England and for Encouraging Such Whose Hearts God Shall Move to Join with Them in It*. London, 1629.

Winthrop, John, Jr. "Concerning Some Natural Curiosities of Those Parts, Especially a Very Strange and Very Curiously Contrived Fish, Sent for the Repository of the R. Society." *Philosophical Transactions* (1670), 1151–1153.

Wood, William. *New England's Prospect: A True, Lively, and Experimental Description of that Part of America Commonly Called New England*. London, 1634.

Woodward, Carl R. *Ploughs and Politicks: Charles Read of New Jersey and His Notes on Agriculture, 1715–1774*. New Brunswick, NJ: Rutgers University Press, 1941.

Wynne, John. *General History of the British Empire in North America*. London, 1770.

Young, Arthur. *The Farmer's Letters to the People of England*. London, 1768.

—— *Observations on the Present State of the Wastelands of Great Britain: Published on the Occasion of the Establishment of a New Colony on the Ohio*. London, 1773.

—— *Annals of Agriculture and Other Useful Arts*. 45 vols. London, 1784–1815.

UNPUBLISHED THESES

Beanlands, Sara J. "Annotated Edition of Rev. Dr. Andrew Brown's Manuscript: 'Removal of the French inhabitants of Nova Scotia by Lieut. Governor Lawrence & His Majesty's Council in October 1755.'" M.A., Saint Mary's University, 2010.

Hill, Michael R. "Temperateness, Temperance, and the Tropics: Climate and Morality in the English Atlantic World, 1555–1705." Ph.D., Georgetown University, 2013.

Picart, Lennox O'Riley. "The Trelawny Maroons and Sir John Wentworth: The Struggle to Maintain Their Culture, 1796–1800." M.A., University of New Brunswick, 1993.

Wickman, Thomas M. "Snowshoe Country: Indians, Colonists, and Winter Spaces of Power in the Northeast, 1620–1727." Ph.D., Harvard University, 2012.

PUBLISHED SECONDARY WORKS

Akita, Shigeru. *Gentlemanly Capitalism, Imperialism, and Global History*. New York: Palgrave, 2002.

Akerman, James R., ed. *The Imperial Map: Cartography and the Mastery of Empire*. Chicago: University of Chicago Press, 2009.

Alexander, Charles P. "The Presidential Range of New Hampshire as a Biological Environment." *American Midland Naturalist* 24(1) (1940), 104–132.

Alsop, James D. "Samuel Vetch's "Canada Survey'd": The Formation of a Colonial Strategy, 1706–1711." *Acadiensis* 12(1) (Autumn 1982), 39–58

—— "The Age of the Projectors: British Imperial Strategy in the North Atlantic in the War of Spanish Succession." *Acadiensis* 21(1) (Autumn 1991), 30–53.

Anderson, David G. "Climate and Culture Change in Prehistoric and Early Historic Eastern North America." *Archaeology of Eastern North America* 29 (2001), 143–186.

Anderson, Virginia D. *New England's Generation: The Great Migration and the Formation of Society and Culture in the Seventeenth Century.* New York: Cambridge University Press, 1991.

—— *Creatures of Empire: How Domestic Animals Transformed Early America.* New York: Oxford University Press, 2004.

Anderson, Warwick. "Climates of Opinion: Acclimatization in Nineteenth-Century France and England." *Victorian Studies* 35(2) (Winter 1992), 135–157.

—— *The Cultivation of Whiteness: Science, Health, and Racial Destiny in Australia.* Melbourne: Melbourne University Press, 2005.

Appalachian Mountain Club. *AMC White Mountain Guide*, 25th ed. Boston: AMC Books, 1992 [1907].

Armitage, David. "Greater Britain: A Useful Category of Historical Analysis?" *AHR* (April 1999), 427–445.

—— *Ideological Origins of the British Empire.* New York: Cambridge University Press, 2000.

—— *Foundations of Modern International Thought.* New York: Cambridge University Press, 2013.

Armitage, David and Michael Braddick, eds. *The British Atlantic World 1500–1800.* New York: Palgrave, 2002.

Armstrong, M.W. "Neutrality and Religion in Revolutionary Nova Scotia." *NEQ* 19 (1946), 52–62.

Ash, Eric H., ed. "Expertise: Practical Knowledge and the Early Modern State." *Osiris* 25 (2010).

Ayers, Edward L. and Peter S. Onuf, eds. *All Over the Map: Rethinking American Regions.* Baltimore: Johns Hopkins University Press, 1996.

Bailyn, Bernard. *The New England Merchants in the Seventeenth Century.* Cambridge, MA: Harvard University Press, 1955.

—— *Voyagers to the West: A Passage in the Peopling of America on the Eve of the Revolution.* New York: Vintage, 1986.

Baker, Emerson W. et al., eds. *American Beginnings: Exploration, Culture, and Cartography in the Land of Norumbega.* Lincoln: University of Nebraska Press, 1994.

Banks, R.E.R., ed. *Sir Joseph Banks: A Global Perspective.* Richmond, VA: Royal Botanic Gardens, 1994.

Barnett, Lydia. "The Theology of Climate Change: Sin as Agency in the Enlightenment's Anthropocene." *Environmental History* 20 (2015), 217–237.

Barrell, John. *The Dark Side of the Landscape: The Rural Poor in English Painting, 1730–1840.* New York: Cambridge University Press, 1980.

Barton, Gregory A. *Empire Forestry and the Origins of Environmentalism.* New York: Cambridge University Press, 2002.

Bashford, Alison. *Global Population: History, Geopolitics, and Life on Earth.* New York: Columbia University Press, 2014.

Bashford, Alison and Carolyn Strange. *Griffith Taylor: Visionary Environmentalist Explorer.* Canberra: National Library of Australia, 2008.

Bayly, C. A. *Imperial Meridian: The British Empire and the World, 1780–1830.* New York: Longman, 1989.

—— *The Birth of the Modern World, 1780–1914: Global Connections and Comparisons.* Malden, MA: Blackwell, 2004.

Beinart, William, and Lotte Hughes. *Environment and Empire.* New York: Oxford University Press, 2007.

Belich, James. *Replenishing the Earth: The Settler Revolution and the Rise of the Anglo-World, 1783–1939.* New York: Oxford University Press, 2009.

Bell, Winthrop P. *The 'Foreign Protestants' and the Settlement of Nova Scotia.* Toronto: University of Toronto Press, 1961.

Benton, Lauren. "Legal Spaces of Empire: Piracy and the Origins of Ocean Regionalism." *Comparative Studies in Society and History* 47(4) (October 2005), 700–724.

—— "Spatial Geographies of Empire." *Itinerario* 30(3) (2006), 19–34.

—— *A Search for Sovereignty: Law and Geography in European Empires, 1450–1900.* New York: Cambridge University Press, 2010.

Bermingham, Ann. *Landscape and Ideology: The English Rustic Tradition, 1740–1860.* Berkeley: University of California Press, 1986.

Bidwell, Percy. "The Agricultural Revolution in New England." *AHR* 26(4) (July 1921), 683–702.

Bidwell, Percy W. and John I. Falconer. *The History of Agriculture in the Northern United States, 1620–1860.* Washington, D.C.: The Carnegie Institution of Washington, 1925.

Bitterman, Rusty. *Rural Protest on Prince Edward Island: From British Colonization to the Escheat Movement.* Toronto: University of Toronto Press, 2006.

Boud, R.C. "Scottish Agricultural Improvement Societies, 1723–1835." *Review of Scottish Culture* 1 (1984), 70–90

Bowen, H.V. *Elites, Enterprise, and the Making of the British Overseas Empire, 1688–1775.* New York: St. Martin's, 1996.

Bowler, Peter J. *The Norton History of the Environmental Sciences.* New York: Norton, 1992.

Bradley, Raymond S. and Philip D. Jones, eds. *Climate Since A.D. 1500.* New York: Routledge, 1995.

Braidwood, Stephen J. *Black Poor and White Philanthropists: London's Blacks and the Foundation of the Sierra Leone Settlement, 1786–1791.* Liverpool: Liverpool University Press, 1994.

Brewer, John. *Pleasures of the Imagination: English Culture in the Eighteenth Century.* Chicago: University of Chicago Press, 2000.

Bristow, Peggy, ed. *We're Rooted Here and They Can't Pull Us Up: Essays in African Canadian Women's History.* Toronto: University of Toronto Press, 1994.

Brown, Christopher L. *Moral Capital: Foundations of British Abolitionism.* Chapel Hill: University of North Carolina Press, 2006.

Brown, Christopher L. and Philip D. Morgan, *Arming Slaves: From Classical Times to the Modern Age.* New Haven, CT: Yale University Press, 2006.

Brown, Richard. *Knowledge is Power: The Diffusion of Information in Early America, 1700-1865.* New York: Oxford University Press, 1989.

Brown, Sanford C. *Benjamin Thompson, Count Rumford.* Cambridge, MA: MIT Press, 1979.

Browne, Janet. *The Secular Ark: Studies in the History of Biogeography.* New Haven, CT: Yale University Press, 1983.

Buckner, Philip and David Frank, eds. *Atlantic Canada Before Confederation: The Acadiensis Reader.* Vol. 1. Fredericton, New Brunswick: Acadiensis Press, 1985.

Buckner, Philip and John G. Reid, eds. *The Atlantic Region to Confederation: A History.* Toronto: University of Toronto Press, 1998.

Bumsted, J.M. *The People's Clearance: Highland Emigration to British North America, 1770–1815.* Winnipeg: University of Manitoba Press, 1982.

—— *Land, Settlement, and Politics on Eighteenth-Century Prince Edward Island.* Montreal: McGill-Queen's University Press, 1987.

Bushman, Richard L. *From Puritan to Yankee: Character and the Social Order in Connecticut, 1690–1765.* Cambridge, MA: Harvard University Press, 1967.

—— *The Refinement of America: Persons, Houses, Cities.* New York: Knopf, 1992.

—— "Markets and Composite Farms in Early America." *WMQ* 55 (1998), 351–374.

Brymer, Douglas. "The Jamaica Maroons—How They Came to Nova Scotia—How They Left It." *Foreign and Commonwealth Office Collection.* London,1895.

Byrd, Alexander X. *Captives and Voyagers: Black Migrants Across the Eighteenth-Century British Atlantic World.* Baton Rouge: Louisiana State University Press, 2008.

Cahill, Barry. "Habeas Corpus and Slavery in Nova Scotia: *R v. Hecht, ex parte Rachel,* 1798." *University of New Brunswick Law Journal* 44 (1995), 179–209.

Caldwell, Douglas R. and Russell S. Harmon, eds. *Studies in Military Geography and Geology.* Norwell: Kluwer, 2004.

Calhoon, Robert, Timothy Barnes, and George Rawlyk, eds. *Loyalists and Community in North America.* Westport, CT: Greenwood Press, 1994.

Callison, Candis. *How Climate Change Comes to Matter: The Communal Life of Facts.* Durham, NC: Duke University Press, 2014.

Campbell, Mavis, ed. *Back to Africa: George Ross and the Maroons from Nova Scotia to Sierra Leone.* Trenton, NJ: Africa World Press, 1993.

Cañizares-Esguerra, Jorge. *How to Write the History of the New World: Histories, Epistemologies, and Identities in the Eighteenth-Century Atlantic World*. Palo Alto, CA: Stanford University Press, 2001.

Canny, Nicholas. "The Ideology of English Colonization: From Ireland to America." *WMQ* 30 (1973), 575–598.

—— *Europeans on the Move: Studies on European Migration, 1500–1800*. New York: Oxford University Press, 1994.

Cantor, Geoffrey. *Michael Faraday: Sandemanian and Scientist: A Study of Science and Religion in the Nineteenth Century*. London: Macmillan, 1991.

Caroe, Gwendy. *The Royal Institution: An Informal History*. London: Jon Murray, 1985.

Carrier, Lyman. "Dr. John Mitchell: Naturalist, Cartographer, and Historian." *Annual Report of the American Historical Association* I. Washington, D.C., 1918.

Cash, Philip. *Dr. Benjamin Waterhouse: A Life in Medicine and Public Service, 1754–1846*. Sagamore Beach, MA: Boston Medical Library & Science History Publications, 2006.

Cassidy, David C. "Meteorology in Mannheim: The Palatine Meteorological Society, 1780-1795." *Sudhoffs Archiv* 69(1) (1985), 8–25.

Cayton, Andrew R. L. *The Frontier Republic: Ideology and Politics in the Ohio Country, 1780–1825*. Kent, OH: Kent Sate University Press, 1986.

Chakrabarty, Dipesh."The Climate of History: Four Theses." *Critical Inquiry* 35 (Winter 2009), 197–222.

—— "Climate and Capital: On Conjoined Histories." *Critical Inquiry* 41(1) (Autumn 2014), 1–23.

Chaplin, Joyce E. *An Anxious Pursuit: Agricultural Innovation and Modernity in the Lower South, 1730–1815*. Chapel Hill: University of North Carolina Press, 1993.

—— "Natural Philosophy and an Early Racial Idiom in North America: Comparing English and Indian Bodies." *WMQ* 54(1) (1997), 229–252.

—— *Subject Matter: Technology, the Body, and Science on the Anglo-American Frontier, 1500-1676*. Cambridge, MA: Harvard University Press, 2001.

—— *The First Scientific American: Benjamin Franklin and the Pursuit of Genius*. New York: Basic Books, 2006.

Choquette, Leslie. *Frenchmen into Peasants: Modernity and Tradition in the Peopling of French Canada*. Cambridge, MA: Harvard University Press, 1997.

Christopher, Emma. "A "Disgrace to the very Colour": Perceptions of Blackness and Whiteness in the Founding of Sierra Leone and Botany Bay." *Journal of Colonialism and Colonial History* 9(3) (2011), online.

Clark, Charles, James Leamon, and Karen Bowden, eds. *Maine in the Early Republic: From Revolution to Statehood*. Hanover, NH: University Press of New England, 1988.

Clark, Christopher. *The Roots of Rural Capitalism: Western Massachusetts, 1780–1860*. Ithaca, NY: Cornell University Press, 1990.

Clark, William, Jan Golinski, and Simon Schaffer. *The Sciences in Enlightened Europe*. Chicago: University of Chicago Press, 1999

Coates, Colin. *The Metamorphoses of Landscape and Community in Early Quebec*. Montreal: McGill-Queen's University Press, 2000.

Cohen, Patricia Cline. *A Calculating People: The Spread of Numeracy in Early America*. New York: Routledge, 1999 [1982].

Coleman, Deirdre. *Romantic Colonization and British Anti-Slavery*. New York: Cambridge University Press, 2005.

Colley, Linda. *Britons: Forging the Nation, 1707–1837*. New Haven, CT: Yale University Press, 1995.

—— *The Significance of the Frontier in British History*. Austin: University of Texas Press, 1995.

Conforti, Joseph A. *Imagining New England: Explorations of Regional Identity from the Pilgrims to the Mid-Twentieth Century*. Chapel Hill: University of North Carolina Press, 2001.

Conrad, Margaret, ed. *They Planted Well: New England Planters in Maritime Canada*. Fredericton, New Brunswick: Acadiensis Press, 1988.

Conrad, Margaret, Toni Laidlaw, and Donna Smith, eds. *No Place Like Home: Diaries and Letters of Nova Scotia Women, 1771–1938*. Halifax, NS: Formac Publishing, 1988.

Cook, Sherburne F. *The Indian Population of New England in the Seventeenth Century*. Berkeley: University of California Press, 1976.

Cormack, Lesley. "'Good Fences Make Good Neighbors': Geography as Self-Definition in Early Modern England." *Isis* 82(4) (1991), 639–661.

Craig, Beatrice. *Backwoods Consumers and Homespun Capitalists: The Rise of Market Culture in Eastern Canada*. Toronto: University of Toronto Press, 2009.

Crawford, Rachel. *Poetry, Enclosure, and the Vernacular Landscape, 1700–1830*. New York: Cambridge University Press, 2002.

Cronon, William. *Changes in the Land: Indians, Colonists and the Ecology of New England*. New York: Hill and Wang, 2003 [1983].

Crosby, Alfred W. *Ecological Imperialism: The Biological Expansion of Europe, 900– 1900*. New York: Cambridge University Press, 1986.

Crowley, John E. "'Taken on the Spot': The Visual Appropriation of New France for the Global British Landscape." *Canadian Historical Review* 86(1) (March 2005), 1–28.

—— *The Invention of Comfort: Sensibilities and Design in Early Modern Britain and Early America*. Baltimore: Johns Hopkins University Press, 2001.

Curran, Andrew S. *The Anatomy of Blackness: Science and Slavery in an Age of Enlightenment*. Baltimore: Johns Hopkins University Press, 2011.

Curtin, Philip D. "'The White Man's Grave': Image and Reality, 1780–1850." *Journal of British Studies* 1(1) (1961), 94–110.

Cuthbertson, Brian. *The Loyalist Governor: Biography of Sir John Wentworth*. Halifax: Pergamon Press, 1983.

Davidson, Peter. *The Idea of the North*. London: Reaktion Books, 2005.

Davis, Robert Scott. "Richard Oswald as 'An American': How a Frontier South Carolina Plantation Identifies the Anonymous Author of *American Husbandry* and a Forgotten Founding Father of the United States." *Journal of Backcountry Studies* 8(1) (Spring 2014), 19–34.

Delbourgo, James. *A Most Amazing Scene of Wonders: Electricity and Enlightenment in Early America*. Cambridge, MA: Harvard University Press, 2006.

—— "The Newtonian Slave Body: Racial Enlightenment in the Atlantic World." *Atlantic Studies* 9(2) (2012), 185–207.

Demeritt, David. "Agriculture, Climate, and Cultural Adaptation in the Prehistoric Northeast." *Archaeology of Eastern North America* 19 (Fall 1991), 183–202.

—— "Representing the 'True' St. Croix: Knowledge and Power in the Partition of the Northeast." *WMQ* 54(3) (July 1997), 515–548.

Demos, John. *A Little Commonwealth: Family Life in Plymouth Colony*. New York: Oxford University Press, 2000 [1970].

Depatie, Sylvie. "Jardins et vergers à Montréal au xviii-e siecle." in Depatie et al., *Habitants et Marchands: Vingt ans après: Lectures de l'histoire des XVIIe et XVIIIe siècles canadiens; Twenty Years Later: Reading the History of 17th and 18th-century Canada*. Montreal: McGill-Queen's University Press, 1998.

Dettelbach, Michael. "'A Kind of Linnaean Being': Forster and Eighteenth-Century Natural History," in Nicholas Thomas, ed. *Observations Made During a Voyage Round the World*. Manoa: University of Hawai'i Press, 1996, lv-lxxiv.

DeVos, Paula S. "Natural History and the Pursuit of Empire in Eighteenth-Century Spain." Eighteenth-Century Studies 40(2) (Winter 2007), 209–239.

Dew, Nicholas and James Delbourgo, eds. *Science and Empire in the Atlantic World*. New York: Routledge, 2008.

Dice, Lee R. *The Biotic Provinces of North America*. Ann Arbor: University of Michigan Press, 1943.

Dickson, David. *Old World Colony: Cork and South Munster, 1630–1830*. Cork: Cork University Press, 2005.

Dilke, O.A.W. *Greek & Roman Maps*. Baltimore: Johns Hopkins University Press, 1985.

Dobson, David. *Scottish Emigration to Colonial America, 1607–1785*. Athens: University of Georgia Press, 2004.

Donahue, Brian. *The Great Meadow: Farmers and the Land in Colonial Concord*. New Haven, CT: Yale University Press, 2004.

Donovan, Kenneth, ed. *The Island: New Perspectives on Cape Breton's History, 1713–1990*. Sydney: University College of Cape Breton Press, 1990.

—— "Slaves in Île Royale, 1713–1758." *French Colonial History* 5 (2004), 25–42.

Doolittle, William E. *Cultivated Landscapes of Native North America*. New York: Oxford University Press, 2000.

Douglas, Starr. "The Making of Scientific Knowledge in an Age of Slavery: Henry Smeathman, Sierra Leone and Natural History." *Journal of Colonialism and Colonial History* 9(3) (Winter 2008).

Drayton, Richard. *Nature's Government: Science, Imperial Britain, and the 'Improvement' of the World*. New Haven, CT: Yale University Press, 2000.

Dunlap, Thomas R. *Nature and the English Diaspora: Environment and History in the United States, Canada, Australia, and New Zealand*. New York: Cambridge University Press, 1999.

Edelson, S. Max. *Plantation Enterprise in Colonial South Carolina*. Cambridge, MA: Harvard University Press, 2006.

Edwards, Paul N. *A Vast Machine: Computer Models, Climate Data, and the Politics of Global Warming*. Cambridge, MA: MIT Press, 2010.

Egnal, Marc. *New World Economies: The Growth of the Thirteen Colonies and Early Canada*. New York: Oxford University Press, 1998.

"Eighteenth-Century Agriculture: A Symposium." Special issue of *Agricultural History* 1(43) (Jan. 1969).

Eldred, Janet Carey and Peter Mortensen. *Imagining Rhetoric: Composing Women of the Early United States*. Pittsburgh, PA: University of Pittsburgh Press, 2002.

Ellis, Joseph. *His Excellency: George Washington*. New York: Knopf, 2005.

Elsner, Jas and J.P. Rubies, eds. *Voyages and Visions: Towards a Cultural History of Travel*. London: Reaktion Books, 1999.

Evans, G.N.D. *Uncommon Obdurate: The Several Public Careers of J.F.W. DesBarres*. Salem, MA: Peabody Museum, 1969.

Faragher, John Mack. *A Great and Noble Scheme: The Tragic Story of the Expulsion of the French Acadians from their American Homeland*. New York: Norton, 2005.

Farber, Hannah. "The Rise and Fall of the Province of Lygonia, 1643–1658." *The New England Quarterly* 82(3) (Sep. 2009), 490–513.

Faulkner, Alaric. "Gentility on the Frontiers of Acadia, 1635–1674: An Archaeological Perspective." *New England/New France, 1600–1850*. Dublin Seminar for New England Folk Life Annual Proceedings 1989.

Fergus, Claudius. "'Dread of Insurrection': Abolitionism, Security, and Labor in Britain's West Indian Colonies, 1760-1823." *WMQ* 66(4) (Oct. 2009), 757–780.

Fleming, Donald. *Science and Technology in Providence, 1760–1914*. Providence, RI: Brown University Press, 1952.

Fleming, James R. *Historical Perspectives on Climate Change*. New York: Oxford University Press, 1998.

Floyd-Wilson, Mary. *English Ethnicity and Race in Early Modern Drama*. New York: Cambridge University Press, 2003.

Frey, Silvia. *Water from the Rock: Black Resistance in a Revolutionary Age*. Princeton, NJ: Princeton University Press, 1991.

Fritts, H. C. and J. M. Lough. "An Estimate of Average Annual Temperature Variations for North America, 1602 to 1961." *Climatic Change* 7(2) (1985), 203–224.

Frost, Alan. *Sir Joseph Banks and the Transfer of Plants To and From the South Pacific, 1786–1798*. Melbourne: The Colony Press, 1993.

Fyfe, Christopher. *A History of Sierra Leone*. New York: Oxford University Press, 1962.

Games, Alison. *Migration and the Origins of the English Atlantic World, 1500–1800.* Cambridge, MA: Harvard University Press, 1999.

Gascoigne, John. *Science in the Service of Empire: Joseph Banks, the British State, and the Uses of Science in the Age of Revolution*. New York: Cambridge University Press, 1998.

Glacken, Clarence J. *Traces on the Rhodian Shore: Nature and Culture in Western Thought From Ancient Times to the Present.* Berkeley: University of California Press, 1967.

Geggus, David. "The Enigma of Jamaica in the 1790s: New Light on the Causes of Slave Rebellions." *WMQ* 44(2) (April 1987), 274–299.

Gerbi, Antonello. *The Dispute of the New World: The History of a Polemic, 1750–1900.* Rev. ed. translated by Jeremy Moyle. Pittsburgh, PA: University of Pittsburgh Press, 1973.

Gillies, John. *Shakespeare and the Geography of Difference.* New York: Cambridge University Press, 1994.

Gilroy, Marion. *Loyalists and Land Settlement in Nova Scotia.* Halifax: Public Archives of Nova Scotia Publication No. 4, Reprint, 1990 [1937].

Goldsmith, Elizabeth C. and Dena Goodman, eds. *Going Public: Women and Publishing in Early Modern France.* Ithaca, NY: Cornell University Press, 1995.

Goldsmith, Elizabeth C., ed. *Writing the Female Voice: Essays on Epistolary Literature.* Boston: Northeastern University Press, 1989.

Golinski, Jan. *British Weather and the Climate of Enlightenment.* Chicago: University of Chicago Press, 2007.

—— *Science as Public Culture: Chemistry and Enlightenment in Britain, 1760–1820.* New York: Cambridge University Press, 1992.

Gómez, Nicolás Wey. *The Tropics of Empire: Why Columbus Sailed South to the Indies.* Cambridge,MA: MIT Press, 2008.

Gorham, Eville. "Titus Smith, A Pioneer of Plant Ecology in North America." *Ecology* 36 (1955), 116–123.

Gould, Eliga H. *Among the Powers of the Earth: The American Revolution and the Making of a New World Empire.* Cambridge, MA: Harvard University Press, 2012.

Gould, Eliga H. and Peter S. Onuf, eds. *Empire and Nation: The American Revolution in the Atlantic World.* Baltimore: Johns Hopkins University Press, 2005.

Graham, Gerald S. *British Policy and Canada, 1774–1791: A Study in Eighteenth Century Trade Policy.* Westport, CT: Greenwood Press, 1974 [1930].

Grandjean, Katherine A. "New World Tempests: Environment, Scarcity, and the Coming of the Pequot War." *WMQ* 68(1) (Jan. 2011), 75–100.

Grant, John N. *The Maroons in Nova Scotia.* Halifax: Formac, 2002.

Grasso, Christopher. *A Speaking Aristocracy: Transforming Public Discourse in Eighteenth-century Connecticut.* Chapel Hill: University of North Carolina Press, 1999.

Greene, Jack. *Pursuits of Happiness: The Social Development of Early Modern British Colonies and the Formation of American Culture.* Chapel Hill: University of North Carolina Press, 1988.

Greene, John C. "The American Debate on the Negro's Place in Nature, 1780–1815." *Journal of the History of Ideas* 15(3) (June 1954), 384–396.

—— *American Science in the Age of Jefferson*. Ames: Iowa State University Press, 1984.

Griffiths, N.E.S. *From Migrant to Acadian: A North American Border People, 1604–1755*. Montreal: McGill-Queen's University Press, 2005.

Gronim, Sara Stidstone. "Geography and Persuasion: Maps in British Colonial New York." *WMQ* 58(2) (Apr. 2001), 373–402.

Gross, Robert A. *The Minutemen and Their World*. New York: Hill and Wang, 1976.

Grove, Richard. *Green Imperialism: Colonial Expansion, Tropical Island Edens, and the Origins of Environmentalism, 1600–1860*. New York: Cambridge University Press, 1996.

—— *Ecology, Climate, and Empire: Colonialism and Global Environmental History, 1400–1940*. Cambridge, MA: White Horse Press, 1997.

Guha, Ranajit. *A Rule of Property for Bengal: An Essay on the Idea of Permanent Settlement*. New Delhi: Orient Longman, 1982.

Gwyn, Julian. *Excessive Expectations: Maritime Commerce and the Economic Development of Nova Scotia*. Montreal: McGill-Queen's University Press, 1998.

Haefeli, Evan and Kevin Sweeney. *Captors and Captives: The 1704 French and Indian Raid on Deerfield*. Amherst: University of Massachusetts Press, 2003.

Hall, David D. *Worlds of Wonder, Days of Judgment: Popular Religious Belief in Early New England*. New York: Knopf, 1989.

Hancock, David. *Citizens of the World: London Merchants and the Integration of the British Atlantic Community*, 1735–1785. New York: Cambridge University Press, 1995.

Harley, J.B. "New England Cartography and the Native Americans." Paul Laxton, ed.*The New Nature of Maps: Essays in the History of Cartography*.Baltimore: Johns Hopkins University Press, 2001, 169–196.

Harlow, Vincent. *The Founding of the Second British Empire, 1763–1793*. 2 vols. London: Longman, 1952–1964.

Harper, Marjorie and Michael E. Vance, eds. *Myth, Migration, and the Making of Memory: Scotia and Nova Scotia c. 1700–1990*. Halifax: Fernwood, 1999.

Harrington, C. R., ed. *The Year Without a Summer? World Climate in 1816*. Ottawa: Canadian Museum of Nature, 1992.

Harrison, Mark. "'The Tender Frame of Man': Disease, Climate, and Racial Difference in India and the West Indies, 1760–1860." *Bulletin of the History of Medicine* 70(1) (1996), 68–93.

—— "Science and the British Empire." *Isis* 96(1) (March 2005), 56–63.

—— *Climates and Constitutions: Health, Race, Environment and British Imperialism in India 1600–1850*. New York: Oxford University Press: 2003.

—— *Medicine in an Age of Commerce and Empire: Britain and Its Tropical Colonies, 1660–1830*. New York: Oxford University Press, 2011.

Headley, John M. "The Sixteenth-Century Venetian Celebration of the Earth's Total Habitability: The Issue of the Fully Habitable World for Renaissance Europe." *Journal of World History* 8(1) (Spring 1997), 1–27.

Helgerson, Richard. "The Land Speaks: Cartography, Chorography, and Subversion in Renaissance England." *Representations* 16 (Autumn 1986), 50–85.

Henretta, James. "Families and Farms: Mentalite in Early America." *WMQ* 35 (1978), 3–32.

Heuman, Gad, ed. *Out of the House of Bondage: Runaways, Resistance, and Marronage in Africa and the New World*. London: Frank Cass, 1986.

Heyrman, Christine. *Commerce and Culture: The Maritime Communities of Colonial Massachusetts, 1690–1750*. New York: Norton, 1984.

Hindle, Brooke. *The Pursuit of Science in Revolutionary America*. Chapel Hill: University of North Carolina Press, 1956.

Hodges, Graham R., ed. *The Black Loyalist Directory: African American in Exile after the American Revolution*. New York: Garland Publishing in association with the New England Historic Genealogical Society, 1996.

Hodson, Christopher. "'A Bondage So Harsh' Acadian Labor in the French Caribbean, 1763–1766." *Early American Studies: An Interdisciplinary Journal* 5(1) (Spring 2007), 95–131.

—— *The Acadian Diaspora: An Eighteenth-Century History*. New York: Oxford University Press, 2010.

Holbrook, Stewart H. *The Yankee Exodus: An Account of Migration from New England*. New York: Macmillan, 1950.

Hornsby, Stephen J. *Surveyors of Empire: Samuel Holland, J.F.W. DesBarres, and the Making of The Atlantic Neptune*. Montreal: McGill-Queen's University Press, 2011.

Hornsby, Stephen and John Reid, eds. *New England and the Maritime Provinces: Connections and Comparisons*. Montreal: McGill-Queen's University Press, 2005.

Houle, Daniel, Jean-David Moore, and Jean Provencher. "Ice Bridges on the St. Lawrence River as an Index of Winter Severity from 1620 to 1910." *Journal of Climate* 20 (2007), 757–764.

Hoyle, Richard W., ed. *Custom, Improvement, and the Landscape in Early Modern Britain*. Burlington, VT: Ashgate, 2011.

Hubbard Jr., Bill *American Boundaries: The Nation, the States, the Rectangular Survey*. Chicago: University of Chicago Press, 2009.

Hudson, Nicolas. "From 'Nation' to 'Race': The Origin of Racial Classification in Eighteenth-Century Thought." *Eighteenth-Century Studies* 29(3) (Spring 1996), 247–264.

Hunter, Phyllis W. *Purchasing Identity in the Atlantic World: Massachusetts Merchants, 1670–1780*. Ithaca, NY: Cornell University Press, 2001.

Innes, Stephen. *Creating the Commonwealth: The Economic Culture of Puritan New England.* New York: Norton, 1995.

Inwood, Kris, ed. *Farm, Factory, and Fortune: New Studies in the Economic History of the Maritime Provinces.* Fredericton, New Brunswick: Acadiensis Press, 1993.

"Itineraries of Atlantic Science–New Questions, New Approaches, New Directions." Special issue of Atlantic Studies: Global Currents 7(4) (2010).

Jaffee, David. "The Village Enlightenment in New England, 1760–1820." *WMQ* 47(3) (July 1990), 327–346.

—— *People of the Wachusett: Greater New England in History and Memory, 1630–1860.* Ithaca, NY: Cornell University Press, 1999.

Jankovic, Vladimir. *Reading the Skies: A Cultural History of English Weather, 1650–1820.* Chicago: University of Chicago Press, 2000.

Jardine, Nicholas, James Secord, and Emma Spary, eds. *Cultures of Natural History.* New York: Cambridge University Press, 1996.

Jasanoff, Maya. *Liberty's Exiles: American Loyalists in the Revolutionary World.* New York: Knopf, 2011.

Johnson, Richard R. *John Nelson, Merchant Adventurer: A Life between Empires.* New York: Oxford University Press, 1991.

Jones, David S. *Rationalizing Epidemics: Meanings and Uses of American Indian Mortality Since 1600.* Cambridge, MA: Harvard University Press, 2004.

Jonsson, Fredrik Albritton. *Enlightenment's Frontier: The Scottish Highlands and the Origins of Environmentalism.* New Haven, CT: Yale University Press, 2013.

—— "Climate Change and the Retreat of the Atlantic: The Cameralist Context of Pehr Kalm's Voyage to North America, 1748-51." *WMQ* 72(2) (Jan. 2015), 99–126.

Judd, Richard W. *The Untilled Garden: Natural History and the Spirit of Conservation in America, 1740–1840.* New York: Cambridge University Press, 2009.

Kamensky, Jane. "'In These Contrasted Climes, How Chang'd the Scene': Progress, Declension, and Balance in the Landscapes of Timothy Dwight." *The New England Quarterly* 63(1) (March 1990), 80–108.

Kernaghan, Lois. "A Man and His Mistress: J.F.W. DesBarres and Mary Cannon." *Acadiensis* 11(1) (1981–1982), 23–42.

Knittle, William A. *Early Eighteenth-Century Palatine Emigration.* Philadelphia: Dorrance & Co., 1937.

Koerner, Lisbet. *Linnaeus: Nature and Nation.* Cambridge, MA: Harvard University Press, 1999.

Kohlstedt, Sally G., ed. *The Origins of Natural Science in America: The Essays of George B. Goode.* Washington, D.C.: Smithsonian Institution Press, 1991.

Kopytoff, Barbara. "Early Jamaican Political Development." *WMQ* 35(2) (1978), 287–397.

Kulikoff, Allan. "Households and Markets: Towards a New Synthesis of American Agrarian History." *WMQ* 50 (1993), 340–355.

—— *From British Peasants to Colonial American Farmers*. Chapel Hill: University of North Carolina Press, 2000.

Kupperman, Karen O. "The Puzzle of the American Climate in the Early Colonial Period." *AHR* 87(5) (1982), 1262–1289.

—— "Fear of Hot Climates in the Anglo-American Colonial Experience." *WMQ* 41(2) (1984), 213–240.

—— "Climate and Mastery of the Wilderness in Seventeenth-Century New England." *Seventeenth Century New England*. Boston: Colonial Society of Massachusetts, 1984, 3–38

Lamb, Hubert H. *The Changing Climate: Selected Papers*. London: Methuen, 1966.

Lanphear, Kim M. and Dean R. Snow. "European Contact and Indian Depopulation in the Northeast: The Timing of the First Epidemics." *Ethnohistory* 35(1) (Winter 1988), 15–33.

Latour, Bruno. *Science in Action: How to Follow Scientists and Engineers Through Society*. Cambridge, MA: Harvard University Press, 1987.

Lawson, Russell M. *The American Plutarch: Jeremy Belknap and the Historian's Dialogue with the Past*. Westport, CT: Greenwood, 1998.

—— *Passaconaway's Realm: Captain John Evans and the Exploration of Mount Washington*. Hanover,NH: University Press of New England, 2002.

LeBlanc, Ronnie-Gilles, ed. *Du Grand Dérangement à la Déportation: Nouvelles perspectives historiques*. Moncton, New Brunswick: Chaire d'études acadiennes, 2005.

Leder, Lawrence H. *Robert Livingston, 1654–1728 and the Politics of Colonial New York*. Chapel Hill: University of North Carolina Press, 1961.

Lewis, Andrew J. "A Democracy of Facts, An Empire of Reason: Swallow Submersion and Natural History in the Early American Republic." *WMQ* 62(4) (Oct. 2005), 663–696.

Li, Hui Lin. "Luigi Castiglioni as a Pioneer in Plant Geography and Plant Introduction." *Proceedings of the American Philosophical Society* 99(2) (Apr. 1955), 51–56.

Locher, Fabien and Jean-Baptiste Fressoz. "Modernity's Frail Climate: A Climate History of Environmental Reflexivity." *Critical Inquiry* 38(3) (Spring 2012), 579–598.

Loehr, Rodney C. "The Influence of English Agriculture on American Agriculture, 1775–1825." *Agricultural History* 11(1) (Jan. 1937), 3–15.

Lockridge, Kenneth. "Land, Population, and the Evolution of New England Society, 1630–1790." *Past & Present* 39 (Apr. 1968), 62–80.

MacKenzie, John M. and T. M. Devine. *Scotland and the British Empire*. New York: Oxford University Press, 2011.

MacKinnon, Neil. *This Unfriendly Soil: The Loyalist Experience in Nova Scotia, 1783–1791*. Montreal: McGill-Queen's University Press, 1986.

MacLaren, I.S. "The Limits of the Picturesque in British North America." *Journal of Garden History* 1 (Jan.–March 1985), 97–111.

MacLeod, Roy M., ed. "Nature and Empire: Science and the Colonial Enterprise." *Osiris*, 2nd Ser., 15 (2000).

Maier, Pauline. "Popular Uprisings and Civil Authority in Eighteenth-Century America." *WMQ* 27(1) (1970), 4–35.

Main, Gloria L. and Jackson T. Main. "Economic Growth and the Standard of Living in Southern New England." *Journal of Economic History* 48(1) (March 1988), 27–46.

Mancall, Peter C. *Fatal Journey: The Final Expedition of Henry Hudson.* New York: Basic Books, 2009

—— "The Raw and the Cold: Five English Sailors in Sixteenth-Century Nunavut." *WMQ* 3rd Ser. 70(1) (Jan. 2013), 3–40.

Mancke, Elizabeth. *The Fault Lines of Empire: Political Differentiation in Massachusetts and Nova Scotia, 1760-1830.* New York: Routledge, 2005.

Mancke, Elizabeth and Carole Shammas. *The Creation of a British Atlantic World.* Baltimore: Johns Hopkins University Press, 2005.

Marcott, Shaun A. et al. "A Reconstruction of Regional and Global Temperature for the Past 11,300 Years." *Science* 339, 1198 (2013), DOI: 10.1126/science.1228026.

Marshall, P.J., ed. *The Oxford History of the British Empire. The Eighteenth Century.* Vol. 2. New York: Oxford University Press, 1998.

—— *Remaking the British Atlantic: The United States and the British Empire after American Independence.* New York: Oxford University Press, 2012.

Marshall, P.J. and Glyndwr Williams. *The Great Map of Mankind: British Perceptions of the World in the Age of Enlightenment.* London: J.M. Dent, 1982.

Marvin, Mary Vaughan. *Benjamin Vaughan, 1751–1835.* Hallowell, Maine: Hallowell Printing, 1979.

Matson, Cathy, ed. *The Economy of Early America: Historical Perspectives and New Directions.* University Park: Penn State University Press, 2006.

McAdam, Jane. *Climate Change, Forced Migration, and International Law.* New York: Oxford University Press, 2012.

McClellan III, James E. *Colonialism and Science: Saint Domingue in the Old Regime.* Baltimore: Johns Hopkins University Press, 1992.

—— *Science Reorganized: Scientific Societies in the Eighteenth Century.* New York: Columbia University Press, 1985.

McClelland, Peter. *Sowing Modernity: America's First Agricultural Revolution.* Ithaca, NY: Cornell University Press, 1997.

McCorison, Marcus A., ed. "A Daybook from the Office of the Rutland Herald." *Proceedings of the American Antiquarian Society* 76(2) (1966), 293–395.

McCorkle, Barbara B. *New England in Early Printed Maps, 1513–1800: An Illustrated Carto-Bibliography.* Providence, RI: John Carter Brown Library, 2001.

McCormick, Ted. *William Petty: And the Ambitions of Political Arithmetic.* New York: Oxford University Press, 2009.

McCusker John J., and Russell R. Menard. *The Economy of British America, 1607–1789.* Chapel Hill: University of North Carolina Press, 1991.

McLaren, John, ed. *Despotic Dominion: Property Rights in British Settler Societies.* Vancouver: University of British Columbia Press, 2004.

McNeill, John R. *Mosquito Empires: Ecology and War in the Greater Caribbean, 1620–1914.* New York: Cambridge University Press, 2010.

McWilliams, James E. "'To Forward Well-Flavored Productions': The Kitchen Garden in Early New England." *The New England Quarterly* 77(1) (March 2004), 25–50.

—— *Building the Bay Colony: Local Economy and Culture in Early Massachusetts.* Charlottesville: University of Virginia Press, 2007.

Melzer, Sara E. *Colonizer or Colonized: The Hidden Stories of Early Modern French Culture.* Philadelphia: University of Pennsylvania Press, 2012.

Merchant, Carolyn. *Ecological Revolutions: Nature, Gender, and Science in New England.* Chapel Hill: University of North Carolina Press, 1989.

Miller, David P. and Peter H. Reill, eds. *Visions of Empire: Voyages, Botany, and Representations of Nature.* New York: Cambridge University Press, 1996.

Miller, Perry. *The New England Mind: From Colony to Province.* Cambridge, MA: Harvard University Press, 1953.

—— *Errand Into the Wilderness.* Cambridge, MA: Harvard University Press, 1984 (1956).

Minardi, Margot. *Making Slavery History: Abolitionism and the Politics of Memory in Massachusetts.* New York: Oxford University Press, 2011.

Mingay, G. E., ed. *The Agrarian History of England and Wales: 1750–1850.* New York: Cambridge University Press, 1989.

—— *The Agricultural Revolution, 1650–1880.* London: Adam and Charles Black, 1977.

—— *Arthur Young and His Times.* New York: Macmillan, 1975.

Mitchison, Rosalind. *Agricultural Sir John: The Life of Sir John Sinclair of Ulbster, 1754–1835.* London: Geoffrey Bles, 1962.

Monaghan, Jennifer E. *Learning to Read and Write in Colonial America.* Amherst: University of Massachusetts Press, 2005.

Moore, Susan Hardman. *Pilgrims: New World Settlers and the Call of Home.* New Haven, CT: Yale University Press, 2007.

Morgan, Edmund S. *The Gentle Puritan: A Life of Ezra Stiles, 1728–1795.* New York: Norton, 1983.

Morison, Samuel Eliot, ed. *William Bradford's Of Plymouth Plantation, 1620–1647.* New York: Knopf, 1952.

Murray, Craig C. *Benjamin Vaughan: The Life of an Anglo-American Intellectual.* New York: Arno Press, 1982.

Naftel, William D. *Prince Edward's Legacy: The Duke of Kent in Halifax, Romance and Beautiful Buildings.* Halifax: Formac, 2005.

Neale, Erskine. *The Life of Field Marshall, HRH, Edward, the Duke of Kent.* London: Richard Bentley, 1850.

Newell, Margaret Ellen. *From Dependency to Independence: Economic Revolution in Colonial New England.* Ithaca, NY: Cornell University Press, 1998.

Noël, Françoise. *The Christie Seigneuries: Estate Management and Settlement in the Upper Richelieu Valley, 176–1859*. Montreal: McGill-Queen's University Press, 1992.

O'Brian, Patrick. *Joseph Banks: A Life*. London: Harvill Press, 1987.

O'Grady, Brendan. *Exiles and Islanders: The Irish Settlers of Prince Edward Island*. Montreal: McGill-Queen's University Press, 2004.

Ogborn, Miles and Charles W.J. Withers. *Georgian Geographies: Essays on Space, Place, and Landscape in the Eighteenth Century*. New York: Manchester University Press, 2004.

Olson, Alison G. *Making the Empire Work: London and American Interest Groups, 1690–1790*. Cambridge,MA: Harvard University Press, 1992.

O'Malley, Therese. "Appropriation and Adaptation: Early Gardening Literature in America." *Huntington Library Quarterly* (Summer 1992), 401–431.

Oreskes, Naomi. "The Scientific Consensus on Climate Change." *Science* 306 (Dec. 2004), 1686.

Osborne, Michael A. "Acclimatizing the World: A History of the Paradigmatic Colonial Science." *Osiris* 2nd Ser. 15 (2000), 135–151.

Oslund, Karen. "Imagining Iceland: Narratives of Nature and History in the North Atlantic." *British Journal of the History of Science* 35 (2002), 313–334.

Otterness, Philip. *Becoming German: The 1709 Palatine Migration to New York*. Ithaca, NY: Cornell University Press, 2004.

Pagden, Anthony. *Lords of All the World: Ideologies of Empire in Spain. Britain, and France, c. 1500-c. 1800*. New Haven, CT: Yale University Press, 1995.

Parker, Geoffrey. *Global Crisis: War, Climate Change, and Catastrophe in the Seventeenth Century*. New Haven, CT: Yale University Press, 2013.

Parrish, Susan Scott. *American Curiosity: Cultures of Natural History in the Colonial British Atlantic World*. Chapel Hill: University of North Carolina Press, 2006.

Pastore, Christopher L. *Between Land and Sea: The Atlantic Coast and the Transformation of New England*. Cambridge, MA: Harvard University Press, 2014.

Pauly, Philip J. "Fighting the Hessian Fly: American and British Responses to Insect Invasion, 1776–1789." *Environmental History* 7 (July 2002), 377–400.

Peace, Thomas G.M. "Deconstructing the Sauvage/Savage in the Writing of Samuel de Champlain and Captain John Smith." *French Colonial History* 7 (2006), 1–20.

Pedley, Mary. "Map Wars: The Role of Maps in the Nova Scotia/Acadia Boundary Disputes of 1750." *Imago Mundi* 50 (1998), 96–104.

—— *The Commerce of Cartography: Making and Marketing Maps in Eighteenth-Century France and England*. Chicago: University of Chicago Press, 2005.

Perl-Rosenberg, Nathan. "Private Letters and Public Diplomacy: The Adams Network and the Quasi-War, 1797–1798." *Journal of the Early Republic* 31(2) (Summer 2011), 283–311.

Phillips, Mark S. *Society and Sentiment: Genres of Historical Writing in Britain, 1740–1820*. Princeton, NJ: Princeton University Press, 2000.

Phillips, Patricia. *The Scientific Lady: A Social History of Women's Scientific Interests, 1520–1918*. New York: St. Martin's Press, 1990.

Plank, Geoffrey G. *An Unsettled Conquest: The British Campaign Agaist the Peoples of Acadia*. Philadelphia: University of Pennsylvania Press, 2001.

Pratt, Mary Louise. *Imperial Eyes: Travel Writing and Transculturation*. New York: Routledge, 1992.

Price, Richard, ed. *Maroon Societies: Rebel Slave Communities in the Americas*. Baltimore: Johns Hopkins University Press, 1996.

Prince, Sue Ann. *Stuffing Birds, Pressing Plants, Shaping Knowledge: Natural History in North America, 1730–1860*. Philadelphia: American Philosophical Society, 2003.

Rappaport, Rhoda. *When Geologists Were Historians, 1665–1750*. Ithaca, NY: Cornell University Press, 1997.

Rawlyk, George. *Nova Scotia's Massachusetts: A Study of Massachusetts-Nova Scotia Relations, 1630–1784*. Montreal: McGill-Queen's University Press, 1973.

Regis, Pamela. *Describing Early America: Bartram, Jefferson, Crèvecoeur, and the Rhetoric of Natural History*. DeKalb: Northern Illinois University Press, 1992.

Reid, John G. *Acadia, Maine, and New Scotland: Marginal Colonies in the Seventeenth Century*. Toronto: University of Toronto Press, 1981.

—— et al. *The 'Conquest' of Acadia, 1710: Imperial, Colonial, and Aboriginal Constructions*. Toronto: University of Toronto Press, 2003.

—— *Essays on Northeastern North America: Seventeenth and Eighteenth Centuries*. Toronto: University of Toronto Press, 2008.

Richards, John F. *The Unending Frontier: An Environmental History of the Early Modern World*. Berkeley: University of California Press, 2003.

Ritvo, Harriet. *The Animal Estate: The English and Other Creatures in the Victorian Age*. Cambridge, MA: Harvard University Press, 1987.

—— "At the Edge of the Garden: Nature and Domestication in Eighteenth- and Nineteenth-Century Britain." *Huntington Library Quarterly* (Summer 1992), 363–378.

—— "Going Forth and Multiplying: Animal Acclimatization and Invasion." *Environmental History* 17(2) (Apr. 2012), 404–414.

Rivers, Isabel and David L. Wykes, eds. *Joseph Priestley: Scientist, Philosopher, and Theologian*. New York: Oxford University Press, 2008.

Roberts, Justin. *Slavery and the Enlightenment in the British Atlantic, 1750–1807*. New York: Cambridge University Press, 2013.

Rothenberg, Winifred. "The Emergence of a Capital Market in Rural Massachusetts, 1730–1838." *Journal of Economic History* 45(4) (1985), 781–808.

—— *From Market Places to a Market Economy: The Transformation of Rural Massachusetts, 1750–1850*. Chicago: University of Chicago Press, 1992.

Rothschild, Emma. "A Horrible Tragedy in the French Atlantic." *Past & Present* 192 (Aug. 2006), 67–108.

Rudwick, Martin J. S. *Bursting the Limits of Time: The Reconstruction of Geohistory in the Age of Revolution*. Chicago: University of Chicago Press, 2005.

Russell, Howard S. *A Long, Deep Furrow: Three Centuries of Farming in New England.* Hanover, NH: University Press of New England, 1982.

Rutman, Darrett B. "Governor Winthrop's Garden Crop: The Significance of Agriculture in the Early Commerce of Massachusetts Bay." *WMQ* 20(3) (1963), 396–415.

Safier, Neil. "Transformations de la zone torride: Les répertoires de la nature tropicale à l'époque des Lumières." *Annales: Histoire, sciences sociales* 66(1) (Jan.–March 2011),143–172.

Sahlins, Peter. *Boundaries: The Making of France and Spain in the Pyrenees.* Berkeley: University of California Press, 1989.

Salisbury, Neal. *Manitou and Providence: Indians, Europeans, and the Making of New England, 1500–1643.* New York: Oxford University Press, 1991.

Samson, Danny. *The Spirit of Industry and Improvement: Liberal Government and Rural-Industrial Society, Nova Scotia, 1790–1862.* Montreal: McGill-Queen's University Press, 2008.

Sayre, Gordon M. *Les Sauvages Américains: Representations of Native Americans in French and English Colonial Literature.* Chapel Hill: University of North Carolina Press, 1997.

Schama, Simon. *Rough Crossings: Britain, the Slaves, and the American Revolution.* London: BBC Books, 2005.

Schiebinger, Londa and Claudia Swan. *Colonial Botany: Science, Commerce, and Politics in the Early Modern World.* Philadelphia: University of Pennsylvania Press, 2005.

Schmidt, Benjamin. "Mapping An Empire: Cartographic and Colonial Rivalry in Seventeenth-Century Dutch and English North America." *WMQ* 54(3) (July 1997), 549–578.

Scodel, Joshua. *Excess and the Mean in Early Modern English Literature.* Princeton, NJ: Princeton University Press, 2002.

Scott, James C. *Domination and the Arts of Resistance: Hidden Transcripts.* New Haven, CT: Yale University Press, 1992.

Shapin, Steven and Simon Schaffer. *Leviathan and the Air-Pump: Hobbes, Boyle, and the Experimental Life.* Princeton, NJ: Princeton University Press, 1985.

Shenstone, Susan Burgess. *So Obstinately Loyal: James Moody, 1744–1809.* Montreal: McGill-Queen's University Press, 2002.

Sherman, William H. "Stirrings and Searchings: 1500–1720." *The Cambridge Companion to Travel Writing.* Peter Hulme and Tim Youngs, eds. New York: Cambridge University Press, 2002, 29–32

Shields, David. *Civil Tongues and Polite Letters in British America.* Chapel Hill: University of North Carolina Press, 1997.

Shipton, Clifford K. *Biographical Sketches of Those Who Attended Harvard College in the Classes 1751–1755.* Boston: Massachusetts Historical Society, 1965.

Shteir, Ann B. *Cultivating Women, Cultivating Science: Flora's Daughters and Botany in England, 1760-1860.* Baltimore: Johns Hopkins University Press, 1996.

Slack, Paul R. *The Invention of Improvement: Information and Material Progress in Seventeenth-Century England*. New York: Oxford University Press, 2014.

Slonosky, Victoria. "The Meteorological Observations of Jean-Francois Gaultier, Quebec, Canada: 1742-56." *Journal of Climate* 16 (2003), 2232–2247.

Smith, Daniel Scott. "The Demographic History of New England." *Journal of Economic History* 32(1) (March 1972), 165–183.

Smith, John H. *The Perfect Rule of the Christian Religion: A History of Sandemanianism in the Eighteenth Century*. Albany, NY: SUNY Press, 2009.

Snider, Alvin. "Hard Frost: 1684." *Journal for Early Modern Cultural Studies* 8(2) (Fall– Winter, 2008), 8–32.

Sparling, Mary. *Great Expectations: The European Vision of Nova Scotia, 1749–1848*. Halifax: Mount Saint Vincent University Art Gallery, 1980.

Spary, Emma C. *Utopia's Garden: French Natural History from the Old Regime to the Revolution*. Chicago: University of Chicago Press, 2000.

Staller, John E. et al., eds. *Histories of Maize: Multidisciplinary Approaches to the Prehistory, Linguistics, Biogeography, Domestication, and Evolution of Maize*. Boston: Elsevier Academic Press, 2006.

Statt, Daniel. *Foreigners and Englishmen: The Controversy over Immigration and Population, 1690–1760*. Newark, DE: University of Delaware Press, 1995.

Stearns, Raymond P. *Science in the British Colonies of America*. Urbana: University of Illinois Press, 1970.

Steele, Ian K. *The English Atlantic, 1675–1740: An Exploration of Communication and Community*. New York: Oxford University Press, 1986.

Steinberg, Ted. *Nature Incorporated: Industrialization and the Waters of New England*. New York: Cambridge University Press, 1991.

Stern, Philip J. and Carl Wennerlind. *Mercantilism Reimagined: Political Economy in Early Modern Britain and Its Empire*. New York: Oxford University Press, 2014.

Stoll, Steven. *Larding the Lean Earth: Soil and Society in Nineteenth-Century America*. New York: Hill and Wang, 2002.

Story, Ronald. *The Forging of an Aristocracy: Harvard and the Boston Upper Class, 1800–1870*. Middletown,CT: Wesleyan University Press, 1980.

Stowell, Marion B. *Early American Almanacs: The Colonial Weekday Bible*. New York: Burt Franklin, 1977.

Sutherland, Willard. *Taming the Wild Field: Colonization and Empire on the Russian Steppe*. Ithaca, NY: Cornell University Press, 2006.

Taylor, Alan. "'Wasty Ways': Stories of American Settlement." *Environmental History* 3:3 (July 1998), 291–310.

—— *Liberty Men and Great Proprietors: The Revolutionary Settlement on the Maine Frontier, 1760–1820*. Chapel Hill: University of North Carolina Press, 1990.

Teich, Mikulas, Roy Porter, and Bo Gustafsson, eds. *Nature and Society in Historical Context*. New York: Cambridge University Press, 1997.

Thomas, Peter A. "Contrastive Subsistence Strategies and Land Use as Factors for Understanding Indian-White Relations in New England." *Ethnohistory* 23(1) (Winter 1976), 1–18.

Thorne, R.G. *The History of Parliament: The House of Commons, 1790–1820*. 5 vols. London: Secker & Warburg, 1986.

Thornton, Tamara P. *Cultivating Gentlemen: The Meaning of Country Life Among the Boston Elite, 1785-1860*. New Haven, CT: Yale University Press, 1989.

Troxler, Carole W. "Re-enslavement of Black Loyalists: Mary Postell in South Carolina, East Florida, and Nova Scotia." *Acadiensis* 37(2) (Summer/Autumn 2008), 70–85.

Ulrich, Laurel Thatcher. *A Midwife's Tale: The Life of Martha Ballard Based on Her Diary, 1785–1812*. New York: Knopf, 1990.

—— *Good Wives: Image and Reality in the Lives of Women in Northern New England*. New York: Vintage, 1991 [1982].

Valencius, Conevery B. *The Health of the Country: How American Settlers Understood Themselves and their Land*. New York: Basic Books, 2002.

Valenze, Deborah. "The Art of Women and the Business of Men: Women's Work and the Dairy Industry c. 1740–1840." *Past & Present* 130 (Feb. 1991), 142–169.

Van Gestel-van het Schip and Paula and Peter van der Krogt, eds. *Mappae antiqua: liber amicorum Günter Schilder: Vriendenboek ter gelegenheid van zijn 65ste verjaardag* [*Essays on the occasion of his 65th birthday*]. 't Goy-Houten: Hes & De Graaf, 2007.

Vetter, Jeremy. "Wallace's *Other* Line: Human Biogeography and Field Practice in the Eastern Colonial Tropics." *Journal of the History of Biology* 39 (2006), 89–123.

Vickers, Daniel. "The Northern Colonies: Economy and Society, 1600–1775." *Cambridge Economic History of the United States*. Robert Gallman and Stanley Engerman, eds. New York: Cambridge University Press, 1996, 209–248.

—— "Those Dammed Shad: Would the River Fisheries of New England Have Survived in the Absence of Industrialization?" *WMQ* 61 (2004), 685–712.

—— *Farmers and Fishermen: Two Centuries of Work in Essex County, Massachusetts, 1630–1850*. Chapel Hill: University of North Caroline Press, 1994.

Vogel, Brant. "The Letter from Dublin: Climate Change, Colonialism, and the Royal Society in the Seventeenth Century." *Osiris* 26(1) (2011), 111–128.

Walker, James S. G. *The Black Loyalist: The Search for a Promised Land in Nova Scotia and Sierra Leone, 1783–1870*. Toronto: University of Toronto Press, 1992.

Wallerstein, Immanuel. *The Modern World-System II: Mercantilism and the Consolidation of the European World Economy, 1600–1750*. Berkeley: University of California Press, 2011 (1980).

Webb, James. *Humanity's Burden: A Global History of Malaria*. New York: Cambridge University Press, 2009.

Weber, Max. *The Protestant Ethic and the Spirit of Capitalism*. New York: Routledge, 2001.

White, Richard. *The Middle Ground: Indians, Empires, and Republics in the Great Lakes Region, 1650–1815*. New York: Cambridge University Press, 1991.

—— *"It's your Misfortune and None of My Own": A History of the American West*. Norman: University of Oklahoma Press, 1991.

White, Sam. *The Climate of Rebellion in the Early Modern Ottoman Empire*. New York: Cambridge University Press, 2011.

—— "The Real Little Ice Age." *Journal of Interdisciplinary History* 44 (2013), 327–352.

Whitfield, Harvey A. *Blacks on the Border: The Black Refugees in British North America, 1815–1860*. Burlington: University of Vermont Press, 2006.

—— "Black Loyalists and Black Slaves in Maritime Canada." *History Compass* 5(6) (Nov. 2007), 1980–1997.

—— "Slavery in English Nova Scotia, 1750–1810," *Journal of the Royal Historical Society of Nova Scotia* 13 (2010).

Wilderson, Paul W. *Governor John Wentworth and the American Revolution: The English Connection*. Hanover, NH: University Press of New England, 1994.

Williams, William H. A. *Tourism, Landscape, and the Irish Character: British Travel Writers in Pre-Famine Ireland*. Madison: University of Wisconsin Press, 2008.

Wilmot, Sarah. *The Business of Improvement: Agriculture and Scientific Culture in Britain, c. 1770–1870*. Bristol, CT: Historical Geography Research Series, No. 24, 1990.

Winks, Robin. *The Relevance of Canadian History: U.S. and Imperial Perspectives*. Toronto: Macmillan, 1979.

—— *The Blacks in Canada*. 2nd ed. Montreal: McGill-Queens University Press, 2003.

Withers, Charles. *Placing the Enlightenment: Thinking Geographically about the Age of Reason*. Chicago: University of Chicago Press, 2007.

—— "Reporting, Mapping, Trusting: Making Geographical Knowledge in the Late Seventeenth Century." *Isis* 90(3) (Sep. 1999), 497–521.

Wolff, Larry. *Inventing Eastern Europe: The Map of Civilization in the Mind of the Enlightenment*. Palo Alto,CA: Stanford University Press, 1997.

Wood, Gillen D'Arcy. *Tambora: The Eruption That Changed the World*. Princeton, NJ: Princeton University Press, 2014.

Wood, Gordon S., *The Radicalism of the American Revolution*. New York: Vintage, 1991.

—— *Revolutionary Characters: What Made the Founders Different*. New York: Penguin, 2006.

Woodward, Walter W. *Prospero's America: John Winthrop Jr., Alchemy, and the Creation of New England Culture, 1606–1676*. Chapel Hill: University of North Carolina Press, 2010.

Wynn, Graeme, "Exciting a Spirit of Emulation Among the 'Plodholes': Agricultural Reform in Pre-Confederation Nova Scotia." *Acadiensis* 20(1) (Autumn 1990), 5–51.

Yirush, Craig. *Settlers, Liberty, and Empire: The Roots of Early American Political Theory, 1675–1775*. New York: Cambridge University Press, 2011.

Zielinski, Gregory A. and Barry D. Keim. *New England Weather, New England Climate.* Hanover, NH: University of New England Press, 2003.

Zilberstein, Anya. "The Natural History of Early Northeastern America: An Inexact Science." Dolly Jørgensen, Finn Arne Jørgensen, and Sara B. Pritchard, eds. *New Natures: Joining Environmental History with Science and Technology Studies.* Pittsburgh, PA: University of Pittsburgh Press, 2013.

—— "Inured to Empire: Wild Rice and Climate Change." *WMQ* 72(1) (Jan. 2015), 125–156.

Index

CPSIA information can be obtained
at www.ICGtesting.com
Printed in the USA
BVHW072109251219
567814BV00002B/17/P